I0499945

Essential Precalculus

A Self-Teaching Guide

Tim Hill

Questing Vole Press

Essential Precalculus: A Self-Teaching Guide
by Tim Hill

Copyright © 2018 by Questing Vole Press. All rights reserved.

Editor: Kevin Debenjak
Proofreader: Diane Yee
Compositor: Kim Frees
Cover: Questing Vole Press

Contents

1 The Real Line and Coordinate Plane

The basic numbers used in algebra are the **real numbers**. Defining the real number system rigorously requires advanced mathematics, but all that's needed for our purposes is a solid grounding and an intuitive grasp. (The adjective "real" was used originally to distinguish these numbers from numbers like $\sqrt{-1}$, which were once thought to be "imaginary" or "unreal".)

The Real Numbers

The real number system contains different types of numbers. The **positive integers** (or **natural numbers**) are the numbers

$$1, 2, 3, 4, 5, \ldots;$$

the **integers** are the numbers

$$\ldots, -3, -2, -1, 0, 1, 2, 3, \ldots;$$

and the **rational numbers** are numbers of the form

$$\frac{m}{n}$$

where m and n are integers and $n \neq 0$. Examples of rational numbers are

$$\frac{1}{2}, -\frac{7}{5}, 0, 3, 6.8622, -5, 4\frac{2}{3}.$$

Note that you can add, subtract, multiply, and divide rational numbers and stay within the system of rational numbers. Rational numbers suffice for all physical measurements of any accuracy (weight, length, area, volume, position, velocity, time, and so on). Algebra, geometry, and calculus, however, require a richer system of numbers that includes irrational numbers. A real number that isn't a rational number is an **irrational number**, for example,

$$\sqrt{2}, \sqrt{3}, \sqrt{2} + \sqrt{3}, \sqrt{5}, \sqrt[3]{5}, \pi, \frac{\pi}{5}, -\sqrt{3}.$$

The decimal expansion of an irrational number ($\pi = 3.14159\ldots$, for example) never repeats or terminates, unlike a rational number.

Recall that $\sqrt{a}$ means the *positive* square root of any positive number a. For example, $\sqrt{4}$ is equal to 2 and not −2, even though $(-2)^2 = 4$. To designate both square roots of 4, we write $\pm\sqrt{4}$. Similarly, $\sqrt[n]{a}$ always means the positive nth root of a.

The types of real numbers form increasingly inclusive subsets.

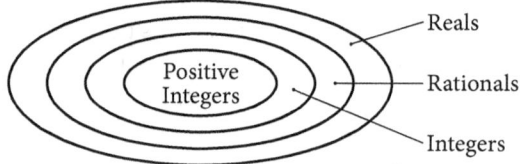

You can also separate the real numbers into the rationals and the irrationals.

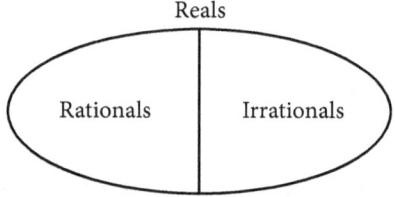

The Real Line

The best way to picture the real number system is by using the **real line** (also called the **number line**). Imagine a horizontal straight line, extending endlessly in both directions. Choose an arbitrary point, called the **origin** or **zero point**, and label it 0. Pick another arbitrary point to the right of 0 and label it 1. The distance between these two points (the **unit distance**) is the measuring scale that associates every real number with a unique point on the line.

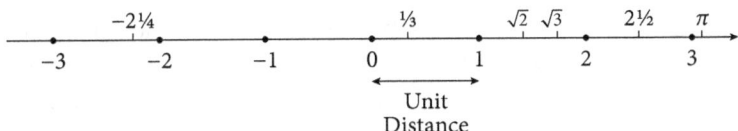

All positive numbers lie to the right of 0, in the positive direction, and all negative numbers lie to the left. (The arrowhead on the right end of the real line indicates the positive direction.) You can draw rational numbers, such as ½, in the obvious way. The irrational numbers also fit into this scheme, so that every real number can be drawn as a point on the line. You can plot irrational numbers on the line by using their decimal expansions: $\sqrt{2} = 1.414\ldots$, $\sqrt{3} = 1.732\ldots$, and $\pi = 3.14159\ldots$. This one-to-one correspondence between all real numbers and all points on the real line establishes these numbers as a coordinate system for the line.

Because this geometric picture plays such a prominent role, geometric terminology is often used when speaking of numbers—thus a number is sometimes called a **point**. It's natural to conflate the distinct concepts of the real number system and the real line, and freely speak of numbers as if they were points on the line and of points on the line as if they were numbers, giving rise to such mixed expressions as "the segment between 3 and 4" and "an irrational point".

Inequalities

The left-to-right linear succession of points on the real line corresponds to the crucial concept of inequalities in the algebra of the real number system. You'll find inequalities playing an increasingly larger role as you progress from precalculus to calculus.

The geometric meaning of the inequality $a < b$ (read "a is less than b") is simply that a lies to the left of b; the equivalent inequality $b > a$ ("b is greater than a") means that b lies to the right of a. A number a is positive or negative according to $a > 0$ or $a < 0$. The main rules used in working with inequalities are:

1. If $a > 0$ and $b < c$, then $ab < ac$.

2. If $a < 0$ and $b < c$, then $ab > ac$.

3. If $a < b$, then $a + c < b + c$ for any number c.

Rules 1 and 2 are usually expressed by saying that an inequality is preserved on multiplication by a positive number, and reversed on multiplication by a negative number; and rule 3 says that an inequality is preserved when any number (positive or negative) is added to both sides. It's often desirable to replace an inequality $a > b$ by the equivalent inequality $a - b > 0$, with rule 3 being used to establish the equivalence.

To state that a is positive or equal to 0, we write $a \geq 0$ ("a is greater than or equal to zero"). Similarly, $a \geq b$ means that $a > b$ or $a = b$. Thus, $3 \geq 2$ and $3 \geq 3$ are both true inequalities.

When you work with inequalities, keep in mind that a product of two or more numbers is zero if and only if one of its factors is zero. If none of its factors are zero, then it's positive or negative depending on whether it has an even or an odd number of negative factors.

Absolute Values

The **absolute value** of a number a is denoted by $|a|$ and defined by

$$|a| = \begin{cases} a & \text{if } a \geq 0, \\ -a & \text{if } a < 0. \end{cases}$$

For example, $|3| = 3$, $|-2| = -(-2) = 2$, and $|0| = 0$. Forming the absolute value leaves positive numbers unchanged and replaces each negative number by the corresponding positive number. The main properties of this operation are

$$|a| \geq 0,$$

$$|ab| = |a||b|,$$

$$|a + b| \leq |a| + |b|.$$

In geometric terms, the absolute value of a number a is simply the distance from the point a to the origin. Similarly, the distance from a to b is $|a - b|$.

To solve an equation such as $|x + 2| = 3$, write it in the form $|x - (-2)| = 3$ and think of it as "the distance from x to -2 is 3". With a mental picture of the real line, it's evident that the solutions are $x = 1$ and $x = -5$. We can also solve this equation by using the fact that $|x + 2| = 3$ means $x + 2 = 3$ or $x + 2 = -3$; the solutions are $x = 1$ and $x = -5$, as before.

Intervals

The sets of real numbers that we'll deal with most frequently are intervals. An **interval** is simply a segment on the real line. If its endpoints are the numbers a and b, then the interval consists of all numbers that lie between a and b. We'll also need to specify whether the endpoints themselves are part of the interval.

Suppose that a and b are numbers, with $a < b$. The **closed interval** from a to b, denoted by $[a, b]$—using brackets—includes its endpoints, and therefore consists of all real numbers x such that $a \leq x \leq b$. Parentheses indicate excluded endpoints. The interval (a, b), with both endpoints excluded, is the **open interval** from a to b; it consists of all x such that $a < x < b$. Sometimes we want to include only one endpoint in an interval. Thus, the intervals denoted by $[a, b)$ and $(a, b]$ are defined by the inequalities $a \leq x < b$ and $a < x \leq b$, respectively. In each of these cases, any number c such that $a < c < b$ is called an **interior point** of the interval.

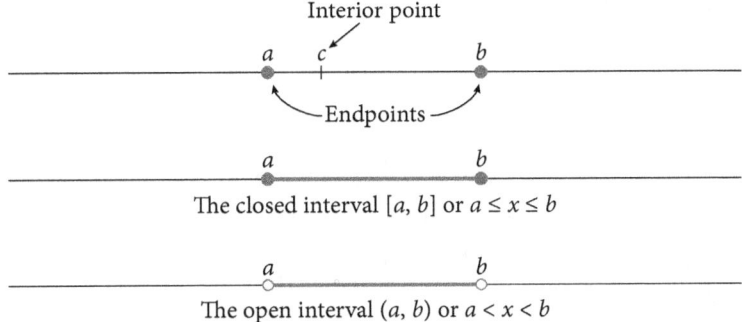

The closed interval $[a, b]$ or $a \leq x \leq b$

The open interval (a, b) or $a < x < b$

Strictly speaking, the notations $a \leq x \leq b$ and $[a, b]$ have different meanings—the first represents a restriction imposed on x, while the second denotes a set—but both designate the same interval. Hence we'll consider them to be equivalent and use them interchangeably, and you should become familiar with both. The geometric meaning of $a \leq x \leq b$ is more easily grasped, however, and so we usually prefer it to $[a, b]$.

A **half-line** is an interval extending to infinity in one direction. The symbol ∞ (read "infinity") is used in designating such an interval. Thus, for any real number a the intervals defined by the inequalities $a < x$ and $x \leq a$ can be written as $a < x < \infty$ and $-\infty < x \leq a$, or equivalently as (a, ∞) and $(-\infty, a]$. Keep in mind that the symbols ∞ and $-\infty$ don't denote real numbers; they're used in this manner only as a convenient way of emphasizing that x is allowed to be arbitrarily large (in either the positive or negative direction). To keep this notation clear in your mind, think of $-\infty$ and ∞ as fictitious numbers located at the left and right "ends" of the real line, as suggested in the following figure. Also, it's sometimes convenient to think of the entire real line itself as an interval, $-\infty < x < \infty$ or $(-\infty, \infty)$.

Sets of numbers described by means of inequalities and absolute values are often intervals. It's clear, for example, that the set of all x such that $|x| < 2$ is the interval $-2 < x < 2$ or $(-2, 2)$.

Example 1.1   Solve the inequality $x^2 - 2 < x$.

Solution   To solve an inequality like this means to find all numbers x for which the inequality is true. Begin by writing it as

$$x^2 - x - 2 < 0,$$

and then write it in the factored form

$$(x + 1)(x - 2) < 0.$$

For this inequality to be true, the two factors must have opposite signs: $x + 1 > 0$ and $x - 2 < 0$, or $x + 1 < 0$ and $x - 2 > 0$. These conditions are equivalent to $x > -1$ and $x < 2$, or $x < -1$ and $x > 2$. The second pair of conditions is clearly impossible. The first pair of conditions means that x lies in the open interval $-1 < x < 2$, and these x's constitute the solution of the given inequality.

The Coordinate Plane

Just as real numbers are used as coordinates for points on a line, pairs of real numbers can be used as coordinates for points in a plane. To do so, we establish a **rectangular coordinate system** in the plane, as follows.

Draw two perpendicular straight lines in the plane, one horizontal and the other vertical, as shown in the following figure. These lines are called the **x-axis** and **y-axis**, respectively, and their point of intersection is called the **origin**. Coordinates are assigned to these axes in the manner described earlier, with the origin as the zero point on both and the same unit of distance measurement on both. The positive x-axis is to the right of the origin and the negative x-axis to the left, as before; and the positive y-axis is above the origin and the negative y-axis below.

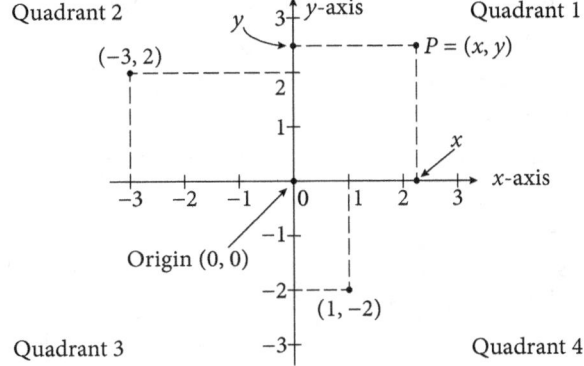

Now consider a point P anywhere in the plane. Draw a line through P parallel to the y-axis, and let x be the coordinate of the point where this line crosses the x-axis. Similarly, draw a line through P parallel to the x-axis, and let y be the coordinate of the point where this line crosses the y-axis. The numbers x and y determined in this way are called the **x-coordinate** and **y-coordinate** of P. When referring to the coordinates of P, it's customary to write them as an ordered pair (x, y) with the x-coordinate written first; we say that P has coordinates (x, y). This correspondence between P and its coordinates establishes a one-to-one correspondence between all points in the plane and all ordered pairs of real numbers. P determines its coordinates uniquely, and by reversing the process we see that each ordered pair of real numbers determines a point P uniquely with these numbers as its coordinates.

As in the case of the real line, it's customary to drop the distinction between a point and its coordinates, and to refer to "the point (x, y)" rather than "the point with coordinates (x, y)". The coordinates x and y of the point P are sometimes called the **abscissa** and **ordinate** of P. Note in particular that points $(x, 0)$ lie on the x-axis, that points $(0, y)$ lie on the y-axis, and that $(0, 0)$ is the origin. Also, the axes divide the plane into four **quadrants,** as shown in the preceding figure, and these quadrants are characterized as follows by the signs of x and y:

first quadrant, $x > 0$ and $y > 0$;
second quadrant, $x < 0$ and $y > 0$;
third quadrant, $x < 0$ and $y < 0$;
fourth quadrant, $x > 0$ and $y < 0$.

When the plane is equipped with the coordinate system described here, it's called the **coordinate plane** or the *xy*-**plane.**

Finally, the notation (x, y) can create ambiguity because it can be used to denote an open interval or an ordered pair, but a careful writer will always make clear from context which is meant.

The Distance Formula

A good deal of our work involves elementary geometric ideas that you learned in earlier mathematics courses (right triangles, similar triangles, circles, spheres, cones, and so on). A major fact of particular importance is the **Pythagorean theorem**: in any right triangle, the sum of the squares of the legs equals the square of the hypotenuse, as shown on the left in the following figure.

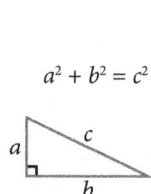

$$a^2 + b^2 = c^2$$

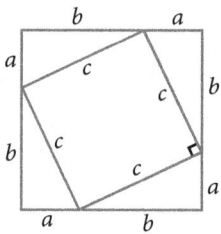

Many proofs of this theorem exist, but the following is simpler than most. Let the legs be a and b and the hypotenuse be c, and arrange four replicas of the triangle in the corners of a square of side $a + b$, as shown on the right in the preceding figure. Thus the area of the large square equals 4 times the area of the triangle plus the area of the small square; that is,

$$(a + b)^2 = 4(\tfrac{1}{2}ab) + c^2.$$

Expand and solve to get $a^2 + b^2 = c^2$, which is the Pythagorean theorem.

We can use this theorem to derive the formula for the distance d between any two points in the coordinate plane. If the points are $P_1 = (x_1, y_1)$ and $P_2 = (x_2, y_2)$, then the segment joining them is the hypotenuse of a right triangle with legs $|x_1 - x_2|$ and $|y_1 - y_2|$.

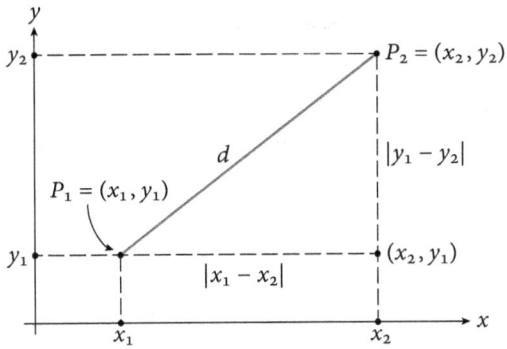

Applying the Pythagorean theorem,

$$d^2 = |x_1 - x_2|^2 + |y_1 - y_2|^2$$
$$= (x_1 - x_2)^2 + (y_1 - y_2)^2,$$

yields the **distance formula,**

$$d = \sqrt{(x_1 - x_2)^2 + (y_1 - y_2)^2}.$$

Example 1.2 Find the distance d between the points $(-4, 3)$ and $(3, -2)$.

Solution Applying the distance formula,

$$d = \sqrt{(-4-3)^2 + (3+2)^2} = \sqrt{74}.$$

Note that in applying the distance formula it doesn't matter in which order the points are taken.

Example 1.3 Find the lengths of the sides of the triangle whose vertices are $P_1 = (-1, -3)$, $P_2 = (5, -1)$, $P_3 = (-2, 10)$.

Solution Applying the distance formula, these lengths are

$$P_1P_2 = \sqrt{(-1-5)^2 + (-3+1)^2} = \sqrt{40} = 2\sqrt{10},$$
$$P_1P_3 = \sqrt{(-1+2)^2 + (-3-10)^2} = \sqrt{170},$$
$$P_2P_3 = \sqrt{(5+2)^2 + (-1-10)^2} = \sqrt{170}.$$

These calculations reveal that the triangle is isosceles, with P_1P_3 and P_2P_3 as the equal sides.

The Midpoint Formulas

It's often useful to know the coordinates of the midpoint of the segment joining two given distinct points. If the given points are $P_1 = (x_1, y_1)$ and $P_2 = (x_2, y_2)$, and if $P = (x, y)$ is the midpoint, then it's clear from the following figure that x is the midpoint of the projection of the segment on the x-axis, and similarly for y.

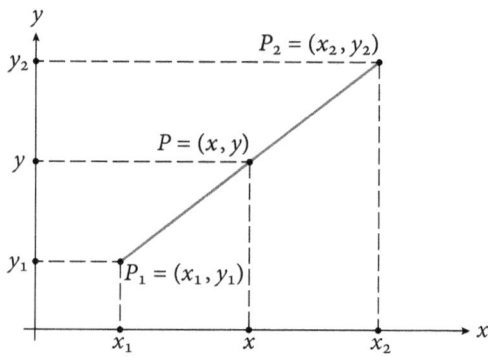

This tells us—examine the figure carefully—that $x = x_1 + \frac{1}{2}(x_2 - x_1)$ and $y = y_1 + \frac{1}{2}(y_2 - y_1)$, so

$$x = \tfrac{1}{2}(x_1 + x_2) \qquad \text{and} \qquad y = \tfrac{1}{2}(y_1 + y_2).$$

Another way to derive these formulas is to notice from the preceding figure that $x - x_1 = x_2 - x$, so $2x = x_1 + x_2$ or $x = \frac{1}{2}(x_1 + x_2)$, with the same argument applying to y. Similarly, if P is a trisection point of the segment joining P_1 and P_2, then its coordinates can be found from the fact that x and y are trisection points of the corresponding segments on the x-axis and y-axis.

Example 1.4 Prove that in any triangle the segment joining the midpoints of two sides is parallel to the third side and half its length.

Solution We already know this fact from elementary geometry, but now we want to prove it by using our methods. We begin by noticing that the triangle can always be positioned as shown in the following figure, with its third side along the positive x-axis and the left endpoint of this side at the origin.

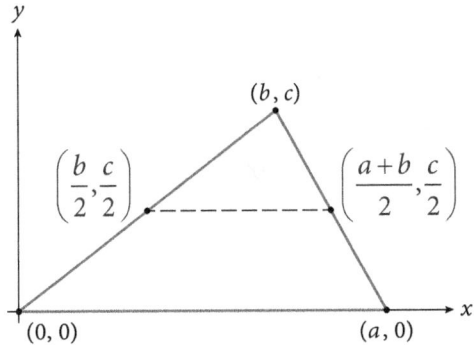

We then insert the midpoints of the other two sides, as shown, and observe that because they have the same y-coordinate, the segment joining them is parallel to the third side lying on the x-axis. The length of this segment is simply the difference between the x-coordinates of its endpoints,

$$\frac{a+b}{2} - \frac{b}{2} = \frac{a}{2},$$

which is half the length of the third side.

Problems

1. Describe the following numbers by using one or more of the terms "integer", "rational", and "irrational":

   (a) −2/3

   (b) 0

   (c) 45/9

   (d) 0.75

   (e) $-\sqrt{49}$

   (f) $1/\pi$

   (g) 9.000...

   (h) $3^{1/2}$

   (i) −20/7

   (j) 94/7

2. Every integer is either even or odd. The even integers are those that are divisible by 2, so n is even if and only if it has the form $n = 2k$ for some integer k. The odd integers are those that have the form $n = 2k + 1$ for some integer k.

   (a) If n is even, prove that n^2 is also even.

   (b) If n is odd, prove that n^2 is also odd.

3. Rewrite the following expressions without using the absolute value symbol:

   (a) $|7 - 18|$

   (b) $|7| - |{-18}|$

   (c) $|\pi - 3|$

   (d) $|3 - \pi|$

   (e) $|x - 5|$ if $x < 5$

   (f) $|x - 5|$ if $x > 5$

   (g) $|x^2 + 10|$

   (h) $|1 - 3x^2|$ if $x \geq 1$

   (i) $|x|/x$ if $x < 0$

   (j) $|x|/x$ if $x > 0$

4. Solve the following inequalities:

   (a) $x(x - 1) > 0$

   (b) $(x - 1)(x + 2) < 0$

   (c) $x^2 + 4x - 21 > 0$

   (d) $2x^2 + x < 3$

   (e) $4x^2 + 10x - 6 < 0$

   (f) $x^2 + 2x + 4 > 0$

5. Recall that $\sqrt{a}$ is a real number if and only if $a \geq 0$, and find the values of x for which each of the following is a real number:

   (a) $\sqrt{4 - x^2}$

   (b) $\sqrt{x^2 - 9}$

   (c) $\dfrac{1}{\sqrt{4 - 3x}}$

   (d) $\dfrac{1}{\sqrt{x^2 - x - 12}}$

6. Find the values of x for which each of the following is positive:

(a) $\dfrac{x}{x^2 + 4}$

(b) $\dfrac{x}{x^2 - 4}$

(c) $\dfrac{x+1}{x-3}$

(d) $\dfrac{x^2-1}{x^2-3x}$

7. State the values of a for which the following inequalities are valid:

(a) $a \le a$

(b) $a < a$

8. If $a \le b$ and $b \le a$, then what conclusion can be drawn about a and b?

9. (a) If $a < b$ is true, then is it also necessarily true that $a \le b$?

(b) If $a \le b$ is true, then is it also necessarily true that $a < b$?

10. State whether each pair of points lies on a horizontal or a vertical line:

(a) $(-2, -5)$, $(-2, 3)$

(b) $(-2, -5)$, $(7, -5)$

(c) $(-3, 4)$, $(6, 4)$

(d) $(2, -11)$, $(2, 5)$

(e) $(2, 2)$, $(-13, 2)$

(f) $(-7, -7)$, $(-7, 7)$

(g) $(3, 5)$, $(3, -2)$

(h) $(-1, -2)$, $(2, -2)$

11. Three vertices of a rectangle are $(-1, 2)$, $(3, -5)$, and $(-1, -5)$. What is the fourth vertex?

12. Find the distance between each pair of points:

(a) $(1, 2)$, $(6, 7)$

(b) $(2, 5)$, $(-1, 3)$

(c) $(-7, 3)$, $(1, -2)$

(d) (a, b), (b, a)

13. Draw a sketch indicating the points (x, y) in the plane for which:
    (a) $x < 2$
    (b) $-1 < y \le 2$
    (c) $0 \le x \le 1$ and $0 \le y \le 1$
    (d) $x = -1$
    (e) $y = 3$
    (f) $x = y$

14. Show that the point $(6, 5)$ lies on the perpendicular bisector of the segment joining the points $(-2, 1)$ and $(2, -3)$.

15. The two points $(2, -2)$ and $(-6, 5)$ are the endpoints of a diameter of a circle. Find the center and radius of the circle.

16. Find the point equidistant from the three points $(-9, 0)$, $(6, 3)$, and $(-5, 6)$.

17. What symmetry statement can be made about the points (a, b) and (b, a)?

18. In an isosceles right triangle, both acute angles are 45°. If the hypotenuse is h, what is the length of each of the other sides?

19. If a and b are positive numbers, prove the inequality $\sqrt{ab} \le \frac{1}{2}(a + b)$ by considering a right triangle inscribed in a semicircle.

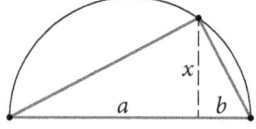

20. Show that if $a \le b$ and $c \le d$, then $a + c \le b + d$. Use this fact to prove that $|a + b| \le |a| + |b|$. (Hint: Begin by noticing that $-|a| \le a \le |a|$ and $-|b| \le b \le |b|$.)

21. If a is a positive rational number, then the **Babylonian method** for calculating the square root of a works as follows. First, choose a rational number that's a reasonable guess at the value of $\sqrt{a}$, and call this initial approximation x_1. Next, divide a by x_1 and average the result with x_1, thereby obtaining a second approximation x_2.

Next, divide a by x_2 and average the result with x_2, obtaining a third approximation x_3. This procedure is expressed by the formula

$$x_{n+1} = \frac{1}{2}\left(x_n + \frac{a}{x_n}\right), \quad n = 1, 2, 3, \dots .$$

Use this method to calculate $\sqrt{2}$, first with $x_1 = 1$ and then with $x_1 = 3/2$.

22. If a and b are real numbers with $a < b$, then show that there exists at least one rational number c such that $a < c < b$, and hence infinitely many. In particular, between any two irrationals there exist an infinite number of rationals.

23. If a and b are irrational, then is $a + b$ necessarily irrational? Is ab?

24. Give another proof of the Pythagorean theorem by using the equations

$$\frac{a}{c} = \frac{e}{a} \quad \text{and} \quad \frac{b}{c} = \frac{d}{b},$$

obtained from similar triangles in the following figure.

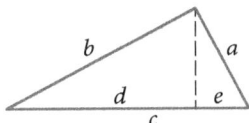

25. If $P_1 = (x_1, y_1)$ and $P_2 = (x_2, y_2)$ are distinct points, and if $P = (x, y)$ is located on the segment joining them in such a position that the ratio of its distance from P_1 to its distance from P_2 is q/p, show that

$$x = \frac{px_1 + qx_2}{p+q} \quad \text{and} \quad y = \frac{py_1 + qy_2}{p+q}.$$

2 Straight Lines

The ideas of this chapter provide valuable tools for proving geometric theorems by using algebraic methods. We use the language of algebra to describe the set of all points that lie on a given straight line. This algebraic description is called the **equation of the line**. First, we must discuss a preliminary concept: the slope of a line.

The Slope of a Line

Every nonvertical straight line has a number associated with it that specifies its direction, called its **slope**, defined as follows. Choose any two distinct points on the line, say $P_1 = (x_1, y_1)$ and $P_2 = (x_2, y_2)$. The slope, denoted by m, is the ratio

$$m = \frac{y_2 - y_1}{x_2 - x_1}. \qquad (1)$$

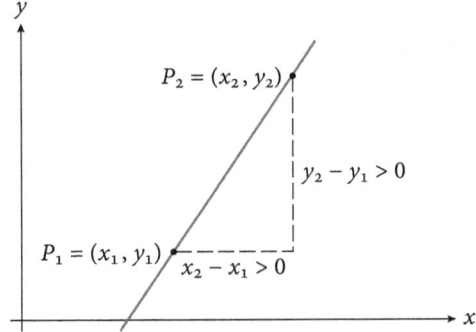

If we reverse the order of subtraction in both numerator and denominator, then the sign of each is changed, so m is unchanged:

$$m = \frac{y_2 - y_1}{x_2 - x_1} = \frac{y_1 - y_2}{x_1 - x_2}.$$

This result shows that the slope can be computed as the difference of the y-coordinates divided by the difference of the x-coordinates—in either order, provided that both differences are formed in the same order. In the preceding figure, where P_2 is placed to the right of P_1 and the line rises to the right, the slope as defined by (1) is simply the ratio of the height to the base in the indicated right triangle. Note that the value of m depends only on the line itself and is the same no matter where the points P_1 and P_2 are located on the line. You can visualize this fact by mentally moving P_1 and P_2 to different positions on the line; this change results in a similar right triangle and therefore leaves the ratio in (1) unchanged.

If we choose the position of P_2 so that $x_2 - x_1 = 1$—that is, if we place P_2 one unit to the right of P_1—then $m = y_2 - y_1$. This tells us that the slope is simply the change in y as a point (x, y) moves along the line in such a way that x increases by one unit. This change in y can be positive, negative, or zero, depending on the direction of the line. We therefore have the following relationships between the sign of m and the indicated directions:

$m > 0,$ line rises to the right;

$m < 0,$ line falls to the right;

$m = 0,$ line horizontal.

Furthermore, the absolute value (page 4) of m is a measure of the steepness of the line. The following figure shows a variety of slopes.

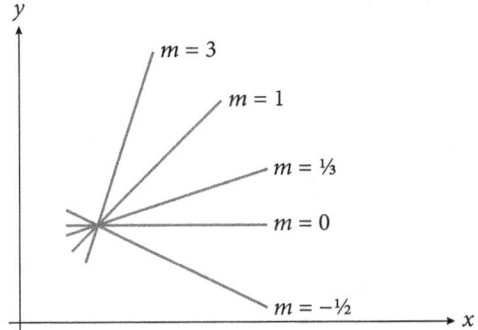

It's evident from (1) that a vertical line has no slope, since the two points have equal x-coordinates and the denominator in (1) is 0—and division by 0 is undefined.

If a line crosses the x-axis, then the angle α from the positive x-direction to the line, measured counterclockwise, is called the **inclination** or the **angle of inclination** of the line. Recall from trigonometry (Chapter 6) that the slope is the tangent of this angle, $m = \tan \alpha$, as shown in the following figure.

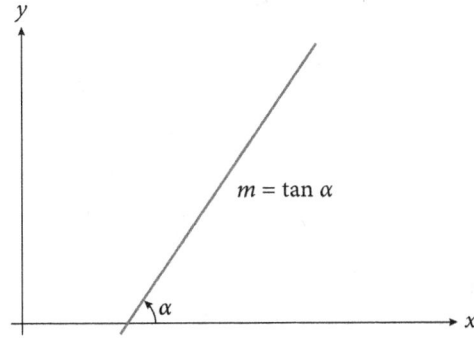

Equations of Lines

On a vertical line, all points have the same x-coordinate. If the line crosses the x-axis at the point $(a, 0)$, then a point (x, y) lies on the line if and only if

$$x = a, \qquad (2)$$

as shown in the following figure. Statement (2), as the equation of the line, tells us that a point (x, y) lies on the line if and only if condition (2) is satisfied.

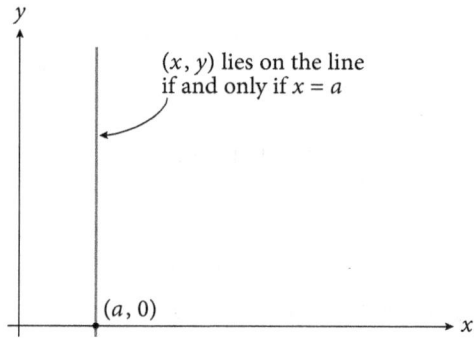

(x, y) lies on the line if and only if $x = a$

$(a, 0)$

Next consider a nonvertical line with a point (x_0, y_0) on it and its slope m, as shown in the following figure.

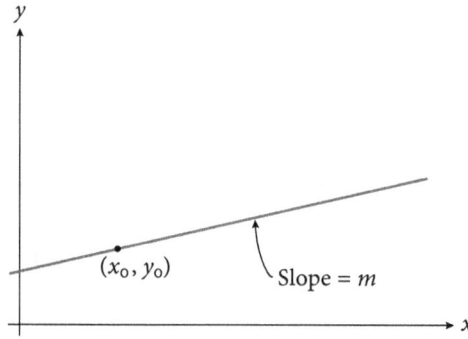

(x_0, y_0)

Slope $= m$

If (x, y) is a point in the plane that doesn't lie on the vertical line through (x_0, y_0), then it's easy to see that this point lies on the given line if and only if the line determined by (x_0, y_0) and (x, y) has the same slope as the given line:

$$\frac{y - y_0}{x - x_0} = m. \qquad (3)$$

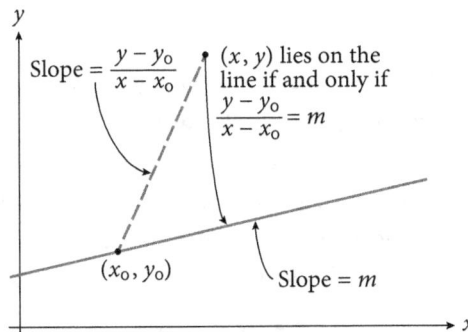

This result would be the equation of our line except for the minor flaw that the coordinates of the point (x_0, y_0)—which is on the line—don't satisfy the equation (they reduce the left side to the meaningless expression 0/0). This flaw is easily removed by writing equation (3) in the form

$$y - y_0 = m(x - x_0). \qquad (4)$$

Nevertheless, we usually prefer the form (3) because its direct connection with the geometric idea illustrated in the preceding figure makes it easy to remember. Either equation (or both) is called the **point-slope equation** of a line, since the line is specified initially by means of a known point on it and its known slope. To grasp the meaning of equation (4), imagine a point (x, y) moving along the given line. As this point moves, its coordinates x and y change; but even though they change, they're bound together by the fixed relationship expressed by equation (4).

If the known point on the line happens to be the point where the line crosses the y-axis, and if this point is denoted by $(0, b)$, then equation (4) becomes $y - b = mx$ or

$$y = mx + b. \qquad (5)$$

The number b is called the **y-intercept** of the line, and (5) is called the **slope-intercept equation** of a line. This form is especially convenient because it tells at a glance the location and direction of a line. For example, if the equation

$$6x - 2y - 4 = 0 \qquad (6)$$

is solved for y, we see that

$$y = 3x - 2. \qquad (7)$$

Comparing (7) with (5) shows instantly that $m = 3$ and $b = -2$, and so (6) and (7) both represent the line that passes through $(0, -2)$ with slope 3, making it easy to sketch the line. It might seem like (6) and (7) are different equations, so that (6) is "an" equation of the line and (7) is "another" equation of the line, but they're merely different forms of a single equation. Many other forms are possible, for example,

$$y + 2 = 3x, \quad x = \tfrac{1}{3}y + \tfrac{2}{3}, \quad 3x - y = 2.$$

It's reasonable to cut through appearances and speak of any one of these as "the" equation of the line.

More generally, every equation of the form

$$Ax + By + C = 0, \qquad (8)$$

where the constants A and B are not both zero, represents a straight line. If $B = 0$, then $A \neq 0$, and the equation can be written as

$$x = -\frac{C}{A},$$

which is clearly the equation of a vertical line. On the other hand, if $B \neq 0$, then

$$y = -\frac{A}{B}x - \frac{C}{B},$$

and this equation has the form (5) with $m = -A/B$ and $b = -C/B$. Equation (8) is somewhat inconvenient for most purposes because its constants aren't directly related to the geometry of the line. Its main merit is that it's capable of representing all lines, with no need to distinguish between the vertical and nonvertical cases. For this reason it's called the **general linear equation**.

Parallel and Perpendicular Lines

Two distinct nonvertical straight lines with slopes m_1 and m_2 are parallel if and only if their slopes are equal:

$$m_1 = m_2.$$

The criterion for perpendicularity is the relation

$$m_1 m_2 = -1. \qquad (9)$$

This fact isn't obvious, but can be proved easily by using similar triangles. Suppose that the lines are perpendicular, as shown in the following figure.

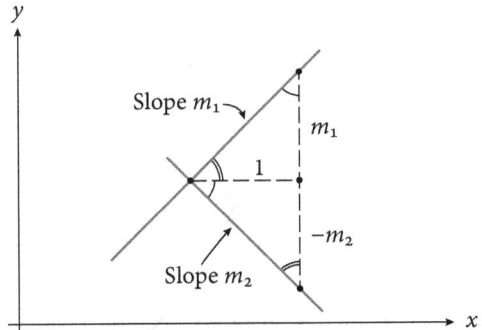

Draw a segment of length 1 to the right from their point of intersection, and from its right endpoint draw vertical segments up and down to the two lines. From the definition of the slopes, the two right triangles formed in this way have sides of the indicated lengths. Because the lines are perpendicular, the indicated angles are equal and the two triangles are similar. This similarity implies that the following ratios of corresponding sides are equal:

$$\frac{m_1}{1} = \frac{1}{-m_2}.$$

This equation is equivalent to (9), so (9) is true when the lines are perpendicular. This reasoning is easily reversed, telling us that if (9) is true, then the lines are perpendicular. Because equation (9) is equivalent to

$$m_1 = -\frac{1}{m_2} \quad \text{and} \quad m_2 = -\frac{1}{m_1},$$

we see that two nonvertical lines are perpendicular if and only if their slopes are negative reciprocals of each other.

Example 2.1 Prove that if the diagonals of a rectangle are perpendicular, then the rectangle is a square.

Solution Position the rectangle as shown in the following figure.

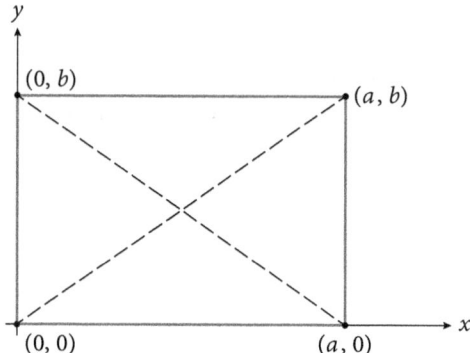

The slopes of the diagonals are clearly b/a and $-b/a$. If these diagonals are perpendicular, then

$$\frac{b}{a} = \frac{a}{b}, \quad a^2 = b^2, \quad a^2 - b^2 = 0, \quad \text{and} \quad (a+b)(a-b) = 0.$$

The last equation implies that $a = b$, so the rectangle is a square.

Problems

1. Plot each pair of points, draw the line they determine, and compute the slope of this line:
   (a) $(-3, 1)$, $(4, -1)$
   (b) $(2, 7)$, $(-1, -1)$
   (c) $(-4, 0)$, $(2, 1)$
   (d) $(-4, 3)$, $(5, -6)$
   (e) $(-5, 2)$, $(7, 2)$
   (f) $(0, -4)$, $(1, 6)$

2. Plot the points $A = (1, 1)$, $B = (-5, -2)$, and $C = (5, 3)$, and use slopes to determine whether all three points lie on a single straight line.

3. Plot the points (–1, –1), (9, 1), (8, 6), and (–2, 4), and show that they are the vertices of a rectangle.

4. Plot each of the following sets of three points, and use slopes to determine in each case whether the points form a right triangle:
   (a) (2, –3), (5, 2), (0, 5)
   (b) (10, –5), (5, 4), (–7, –2)
   (c) (8, 2), (–1, –1), (2, –7)
   (d) (–2, 6), (3, –4), (8, 11)

5. Find the y-intercept of each line in Problem 1 and write the equation in the form $y = mx + b$.

6. Find the equation of the line:
   (a) through (2, –3) with slope –4.
   (b) through (–4, 2) and (3, –1).
   (c) with slope 2/3 and y-intercept –4.
   (d) through (2, –4) and parallel to the x-axis.
   (e) through (1, 6) and parallel to the y-axis.
   (f) through (4, –2) and parallel to $x + 3y = 7$.
   (g) through (5, 3) and perpendicular to $y + 7 = 2x$.
   (h) through (–4, 3) and parallel to the line determined by (–2, –2) and (1, 0).
   (i) that is the perpendicular bisector of the segment joining (1, –1) and (5, 7).
   (j) through (–2, 3) with inclination 135°. (Hint: Use tan θ to find the slope; see Chapter 6.)

7. If a line crosses the x-axis at the point (a, 0), then the number a is called the **x-intercept** of the line. If a line has x-intercept $a \neq 0$ and y-intercept $b \neq 0$, then its equation can be written as

$$\frac{x}{a} + \frac{y}{b} = 1.$$

This equation is called the **intercept form** of the equation of a line. Notice that it's easy to set $y = 0$ and see that the line crosses the x-axis at $x = a$, and to set $x = 0$ and see that the line crosses the y-axis at $y = b$. Put each of the following equations in intercept form and sketch the corresponding line:

(a) $5x + 3y + 15 = 0$

(b) $3x = 8y - 24$

(c) $y = 6 - 6x$

(d) $2x - 3y = 9$

8. Sketch the lines $3x + 4y = 7$ and $x - 2y = 6$, and find their point of intersection. (Hint: Their point of intersection is that point (x, y) whose coordinates satisfy both equations simultaneously.)

9. Let F and C denote temperature in degrees Fahrenheit and degrees Celsius. Find the equation connecting F and C, given that it's linear and that $F = 32$ when $C = 0$, $F = 212$ when $C = 100$.

10. Show that the segments joining the midpoints of adjacent sides of any quadrilateral form a parallelogram.

11. Let $(0, 0)$, $(a, 0)$, and (b, c) be the vertices of an arbitrary triangle placed so that one side lies along the positive x-axis with its left endpoint at the origin. If the square of this side equals the sum of the squares of the other two sides, then use slopes to show that the triangle is a right triangle. Thus, the converse of the Pythagorean theorem (page 8) is also true.

12. If the line determined by two distinct points (x_1, y_1) and (x_2, y_2) is not vertical, and therefore has slope $(y_2 - y_1)/(x_2 - x_1)$, then show that the point-slope form of its equation is the same regardless of which point is used as the given point.

13. The points $(0, 0)$, $(a, 0)$, and (b, c) are the vertices of an arbitrary triangle which is placed in a convenient position relative to the co-ordinate system. Find the equation of the line through each vertex perpendicular to the opposite side, and show algebraically that these three lines intersect at a single point.

14. Show that the distance from a point (x_0, y_0) to a line $Ax + By + C = 0$ is given by $\dfrac{|Ax_0 + By_0 + C|}{\sqrt{A^2 + B^2}}$.

15. If two intersecting straight lines are given, then it's easy to see that the bisectors of the angles formed by these lines are two other straight lines whose points are equidistant from the given lines. Use this fact to find the equations of the bisectors of the angles formed by the lines:

    (a) $3x + 4y - 10 = 0$ and $4x - 3y - 5 = 0$

    (b) $y = 0$ and $y = x$

3

Circles and Parabolas

The coordinate plane or xy-plane is also called the **cartesian plane**, and x and y the cartesian coordinates of the point $P = (x, y)$. The term "cartesian" comes from the name of the French mathematician Descartes, one of the two principal founders of analytic geometry (the other—also French—was Fermat, an even greater mathematician; their names are pronounced "Day-CART" and "Fair-MAY"). The basic idea of this subject is simple: use the tools of algebra to exploit the correspondence between points and their coordinates to study geometric problems, especially the properties of curves. Generally speaking, geometry is visual and intuitive, whereas algebra emphasizes computational tools, and each can serve the other.

You're probably aware that an equation

$$F(x, y) = 0 \qquad (1)$$

usually determines a curve (its *graph*; Chapter 5) that consists of all points $P = (x, y)$ whose coordinates satisfy the given equation. Conversely, a curve defined by some geometric condition can usually be described algebraically by an equation of the form (1). It's clear that straight lines are the simplest curves, and we know from the preceding chapter that straight lines in the coordinate plane correspond to linear equations in x and y. In this chapter, we'll develop algebraic descriptions of several other curves that will be useful as illustrative examples in the next few chapters.

Circles

The distance formula (page 8) is often useful in finding the equation of a curve whose geometric definition depends on one or more distances.

One of the simplest curves of this type is a **circle**, which can be defined as the set of all points at a given distance (the radius) from a given point (the center).

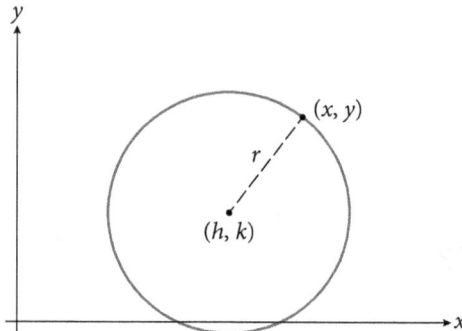

If the center is the point (h, k) and the radius is the positive number r, and if (x, y) is an arbitrary point on the circle, then the defining condition is

$$\sqrt{(x-h)^2 + (y-k)^2} = r.$$

It's convenient to eliminate the radical sign by squaring, which yields

$$(x-h)^2 + (y-k)^2 = r^2. \qquad (2)$$

This is therefore the equation of the circle with center (h, k) and radius r. In particular, if the center is the origin, so that $h = k = 0$, then

$$x^2 + y^2 = r^2$$

is the equation of the circle.

Example 3.1 What is the equation of a circle with center $(-3, 4)$ and radius $\sqrt{10}$.

Solution Applying (2), the circle's equation is

$$(x + 3)^2 + (y - 4)^2 = 10.$$

Note that the coordinates of the center are the numbers *subtracted* from x and y in the parentheses.

Example 3.2 Prove algebraically that an angle inscribed in a semi-circle is always a right angle.

Solution Let the semicircle have radius r and center at the origin $(0, 0)$, so that its equation is $x^2 + y^2 = r^2$ with $y \geq 0$.

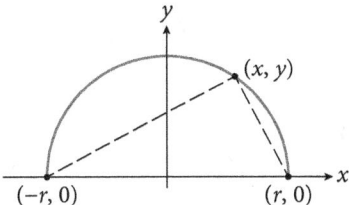

The inscribed angle is a right angle if and only if the product of the slopes of its sides is -1, that is,

$$\frac{y}{x-r} \cdot \frac{y}{x+r} = -1. \qquad (3)$$

This equation is equivalent to $x^2 + y^2 = r^2$, which is true for any point (x, y) on the semicircle, so (3) is true and the angle is a right angle.

Any equation of the form (2) is easy to interpret geometrically. For example,

$$(x - 5)^2 + (y + 2)^2 = 16 \qquad (4)$$

is immediately recognizable as the equation of the circle with center $(5, -2)$ and radius 4, letting us sketch the graph easily. If the equation had been algebraically "simplified", however, then it might have the form

$$x^2 + y^2 - 10x + 4y + 13 = 0. \qquad (5)$$

This equation is an equivalent but scrambled version of (4), and its constants tell us nothing directly about the nature of the graph. To discover what the graph is, we must "unscramble" equation (5) by **completing the square**. To do so, begin by rewriting (5) as

$$(x^2 - 10x + \_\,) + (y^2 + 4y + \_\,) = -13,$$

with the constant term moved to the right and blanks provided for the insertion of suitable constants. When the square of half the coefficient of x is added in the first blank and the square of half the coefficient of y in the second blank, and the same constants are added to the right side to maintain the balance of the equation, we get

$$(x^2 - 10x + 25) + (y^2 + 4y + 4) = -13 + 25 + 4$$

or

$$(x - 5)^2 + (y + 2)^2 = 16. \qquad (6)$$

Note that the form of the equation $(x + a)^2 = x^2 + 2ax + a^2$ is the key to the process of completing the square. The right side is a perfect square—the square of $x + a$—precisely because its constant term is the square of half the coefficient of x.

This same process can be applied to the general equation of the form (5), namely,

$$x^2 + y^2 + Ax + By + C = 0, \qquad (7)$$

but there's little to gain by writing out the details in this general case. However, it's important to notice that if the constant term 13 in (5) is replaced by 29, then (6) becomes

$$(x - 5)^2 + (y + 2)^2 = 0,$$

whose graph is the single point $(5, -2)$. Similarly, if this constant term is replaced by any number greater than 29, then the right-hand side of (6) becomes negative and the graph is empty, meaning that there are no points (x, y) in the plane whose coordinates satisfy the equation. We therefore see that the graph of (7) is sometimes a circle, sometimes a single point, and sometimes empty—depending entirely on the constants A, B, and C.

Parabolas

A **parabola** is a curve consisting of all points that are equally distant from a fixed point F (called the **focus**) and a fixed line d (called the **directrix**). The distance from a point to a line always means the perpendicular distance.

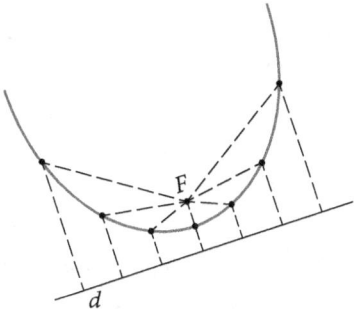

To find a simple equation for a parabola, we place it in the coordinate system as shown in the following figure, with the focus F and directrix d equally far above and below the x-axis.

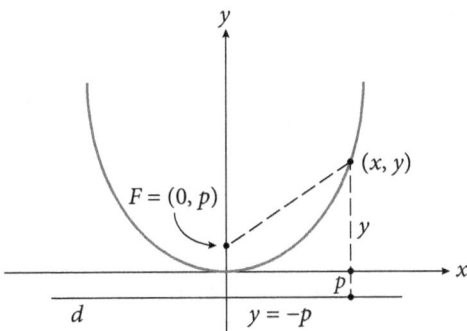

The line through the focus perpendicular to the directrix is called the **axis** of the parabola; this line is the axis of symmetry of the curve, and is the y-axis in the figure. The point on the axis halfway between the focus and the directrix is called the **vertex** of the parabola; in the figure this point is the origin. If (x, y) is an arbitrary point on the parabola, then the condition expressed in the definition is stated algebraically by the equation

$$\sqrt{x^2 + (y - p)^2} = y + p, \qquad (8)$$

Squaring both sides and simplifying yields

$$x^2 + y^2 - 2py + p^2 = y^2 + 2py + p^2$$

or

$$x^2 = 4py. \qquad (9)$$

These steps are reversible, so (8) and (9) are equivalent and (9) is the equation of the parabola whose focus and directrix are located as shown in the preceding figure. Notice in particular that the positive constant p in (9) is the distance from the focus to the vertex, and also from the vertex to the directrix.

If we change the position of the parabola relative to the coordinate axes, then we naturally change its equation. Three other positions are shown in the following figure, each with its corresponding equation and with $p > 0$ in each case. You should verify the correctness of all three equations.

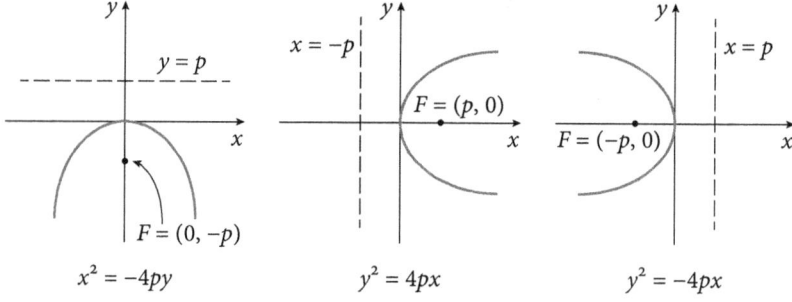

Note that each of these four equations can be put in the form

$$y = ax^2 \qquad (10)$$

or

$$x = ay^2.$$

These forms conceal the constant p, with its geometric significance, but as compensation they're more useful in visualizing the overall appearance of the graph. For example, in (10) the variable x is squared but y is not, telling us that as a point (x, y) moves out along the curve, y increases much faster than x, and so the curve opens in the y-direction—upward or downward, depending on whether a is positive or negative. We can also see that the graph is symmetric with respect to the y-axis, because x is squared, and therefore we get the same number y for any number x and its negative.

Example 3.3   What is the graph of the equation $12x + y^2 = 0$?

Solution   If the given equation is put in the form $y^2 = -12x$ and compared with the equation on the right in the preceding figure, then it's clear that the graph is a parabola with vertex at the origin and opening to the left. Because $4p = 12$ and therefore $p = 3$, the point $(-3, 0)$ is the focus and $x = 3$ is the directrix.

Example 3.4   The graph of $y = 2x^2$ is a parabola with vertex at the origin and opening upward. What is its focus and directrix?

Solution   Rewrite the given equation as $x^2 = \frac{1}{2}y$ and compare it with equation (9). This yields $4p = \frac{1}{2}$, so $p = 1/8$. The focus is therefore $(0, 1/8)$, and the directrix is $y = -1/8$.

We illustrate one last point about parabolas by examining the equation

$$y = x^2 - 4x + 5. \qquad (11)$$

If this equation is written as

$$y - 5 = x^2 - 4x,$$

and if we complete the square on the terms involving x, then the result is

$$y - 1 = (x - 2)^2. \qquad (12)$$

If we now introduce the new variables

$$X = x - 2 \qquad \text{and} \qquad Y = y - 1, \qquad (13)$$

then equation (12) becomes

$$Y = X^2.$$

The graph of this equation is clearly a parabola opening upward with vertex at the origin of the XY coordinate system. By equations (13), the origin in the XY system is the point $(2, 1)$ in the xy system, as shown in the following figure.

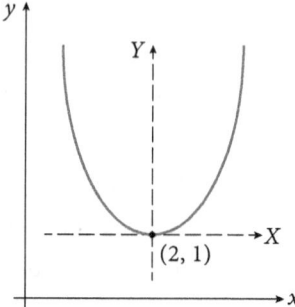

What has happened here is that the coordinate system has been shifted or translated to a new position in the plane, and the axes renamed, and equations (13) express the relation between the coordinates of an arbitrary point with respect to each of the two coordinate systems. In the same way, any equation of the form

$$y = ax^2 + bx + c, \qquad a \neq 0,$$

represents a parabola with vertical axis that's congruent to $y = ax^2$ and opens up or down depending on whether the number a is positive or negative. Similarly, the equation

$$x = ay^2 + by + c, \qquad a \neq 0,$$

represents a parabola with horizontal axis that opens to the right or left depending on whether $a > 0$ or $a < 0$.

In our work up to now we've used the static concept of a curve as a specific set of points or a geometric figure. It's often possible to view a curve as the dynamic path traced by a moving point. For example, a circle is the path of a point that moves in such a way that it maintains a fixed distance from a given point. When this mode of thought is used—with its advantage of greater intuitive vividness—a curve is often called a **locus**. Thus, a parabola is the locus of a point that moves in such a way that it maintains equal distances from a given point and a given line.

Problems

1. Find the equation of the circle with the given point as center and the given number as radius:
   (a) $(4, 6)$, 3
   (b) $(-3, 7)$, $\sqrt{5}$
   (c) $(-5, -9)$, 7
   (d) $(1, -6)$, $\sqrt{2}$
   (e) $(a, 0)$, a
   (f) $(0, a)$, a

2. In each case find the equation of the circle determined by the given conditions:
   (a) Center $(-7, 3)$ and passes through $(4, -1)$.
   (b) The endpoints of a diameter are $(-1, 1)$ and $(7, 9)$.
   (c) Center $(-3, -5)$ and tangent to the line $12x + 5y = 4$. (Hint: See Problem 14 in Chapter 2.)

3. In each of the following, determine the nature of the graph of the given equation by completing the square:
   (a) $x^2 + y^2 - 4x - 4y = 0$
   (b) $x^2 + y^2 - 18x - 14y + 130 = 0$
   (c) $x^2 + y^2 + 8x + 10y + 40 = 0$
   (d) $4x^2 + 4y^2 + 12x - 32y + 37 = 0$
   (e) $x^2 + y^2 - 8x + 12y + 53 = 0$
   (f) $x^2 + y^2 - \sqrt{2}x + \sqrt{2}y + 1 = 0$
   (g) $x^2 + y^2 - 16x + 6y - 48 = 0$

4. The **quadratic formula** for the roots of the quadratic equation $ax^2 + bx + c = 0$ is

   $$x = \frac{-b \pm \sqrt{b^2 - 4ac}}{2a}.$$

   Derive this formula from the equation by dividing through by a, moving the constant term to the right side, and completing the square. Under what circumstances does the equation have distinct real roots, equal real roots, and no real roots?

5. Find the equations of all lines that are tangent to the circle $x^2 + y^2 = 2y$ and pass through the point $(0, 4)$. (Hint: The line $y = mx + 4$ is tangent to the circle if it intersects the circle at only one point.)

6. Find the focus, vertex, and directrix of each of the following parabolas, and sketch the curves:
   (a) $y^2 = 12x$
   (b) $y = 4x^2$
   (c) $2x^2 + 5y = 0$
   (d) $4x + 9y^2 = 0$
   (e) $x = -2y^2$
   (f) $12y = -x^2$
   (g) $16y^2 = x$
   (h) $24x^2 = y$
   (i) $y^2 + 8y - 16x = 16$
   (j) $x^2 + 2x + 29 = 7y$

7. Sketch the parabola and find its equation if it has:
   (a) vertex $(0, 0)$ and focus $(-3, 0)$
   (b) vertex $(0, 0)$ and directrix $y = -1$
   (c) vertex $(0, 0)$ and directrix $x = -2$
   (d) vertex $(0, 0)$ and focus $(0, -1/3)$
   (e) directrix $x = 2$ and focus $(-4, 0)$
   (f) focus $(3, 3)$ and directrix $y = -1$

8. Find the focus, vertex, and directrix of each of the following parabolas, and sketch the curves:
   (a) $y = x^2 + 1$
   (b) $y = (x - 1)^2$
   (c) $y = (x - 1)^2 + 1$
   (d) $y = x^2 - x$

9. Water squirting out of a horizontal nozzle held 4 feet above the ground describes a parabolic curve with the vertex at the nozzle. If the stream of water drops 1 foot in the first 10 feet of horizontal motion, at what horizontal distance from the nozzle will it strike the ground?

10. Show that there is exactly one line with given slope m which is tangent to the parabola $x^2 = 4py$, and find its equation.

11. Prove that the two tangents to a parabola from any point on the directrix are perpendicular.

12. Find the values of b for which the line $y = 3x + b$ intersects the circle $x^2 + y^2 = 4$.

13. Find the equation of the locus of a point $P = (x, y)$ that moves in such a way that:

    (a) Its distance from $(0, 0)$ is twice its distance from $(a, 0)$.

    (b) The product of its distances from $(a, 0)$ and $(-a, 0)$ is a^2 (this curve is called a **lemniscate**).

    In each case, sketch the graph.

14. A point moves in such a way that the ratio of its distances from two fixed points is a constant $k \neq 1$. Show that the locus is a circle.

15. Find the equations of the lines that pass through the point $(1, 3)$ and are tangent to the circle $x^2 + y^2 = 2$. (Hint: See Problem 14 in Chapter 2.)

16. Find the equation of the parabola with focus $(1, 1)$ and directrix $x + y = 0$, and simplify this equation to a form without radicals. (Hint: See Problem 14 in Chapter 2.)

17. Consider all chords with given slope m that have endpoints on the parabola $x^2 = 4py$. Prove that the locus of the midpoints of these chords is a straight line parallel to the y-axis.

4

Functions

The most important concept in all mathematics is that of a function. In every branch of the subject—algebra, geometry, number theory, probability, or any other—functions are the chief objects of investigation. This is especially true of calculus, in which most work concerns constructing tools for studying functions and applying these tools to problems in science and geometry.

The Concept of a Function

What is a function? Briefly, if x and y are two variables that are related in such a way that whenever a permissible numerical value is assigned to x, there is determined one and only one corresponding numerical value for y, then y is called a **function of x**.

Example 4.1   (a) If a rock is dropped from the edge of a cliff, and it falls s feet in t seconds, then s is a function of t. It's known from experiment that (approximately) $s = 16t^2$.

(b) The area A of a circle is a function of its radius r. It's known from geometry that $A = \pi r^2$.

(c) If the manager of a bookstore buys n books from a publisher at \$12 per copy and the shipping charge is \$35, then his cost C for these books is a function of n given by the formula $C = 12n + 35$.

Let's build on the concept of a function by considering an example related directly to our work in the preceding chapter.

Example 4.2 Consider the equation

$$y = x^2$$

and its corresponding graph, which we know is a parabola that opens upward and has its vertex at the origin. In the preceding chapter we thought of this equation as a relation between the variable coordinates of a point (x, y) moving along the curve. We now shift our point of view, and instead think of it as a formula that provides a way to calculate the numerical value of y when the numerical value of x is given. Thus, $y = 1$ when $x = 1$, $y = 4$ when $x = 2$, $y = \frac{1}{4}$ when $x = \frac{1}{2}$, $y = 1$ when $x = -1$, and so on. The value of y is therefore said to **depend on**, or to be a **function of**, the value of x. This dependence can be expressed in functional notation by writing

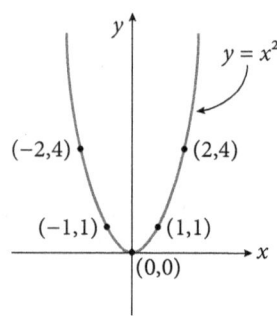

$$y = f(x) \qquad \text{where} \qquad f(x) = x^2.$$

The symbol $f(x)$ is read "f of x", and the letter f represents the rule or process—squaring, in this case—that's applied to any number x to yield the corresponding number y. The numerical examples just given can therefore be written as $f(1) = 1$, $f(2) = 4$, $f(\frac{1}{2}) = \frac{1}{4}$, and $f(-1) = 1$. The meaning of this notation can be further clarified by observing that

$$f(x + 1) = (x + 1)^2 = x^2 + 2x + 1$$

and

$$f(x^3) = (x^3)^2 = x^6;$$

that is, the rule f simply produces the square of whatever quantity follows it in parentheses.

The preceding example suggests the general concept of a function that we'll use in this book, formulated as follows.

Let D be a given set of real numbers. A **function** f defined on D is a formula, or rule, or law of correspondence that assigns a single real number y to each number x in D. The set D of allowed values of x is the **domain** of the function, and the set of corresponding values of y is its **range**. The number y that's assigned to x by the function f is written $f(x)$—so that $y = f(x)$—and is the **value** of f at x. The number x is the **independent variable** because it's free to assume any value in the domain, and y is the **dependent variable** because its numerical value depends on the choice of x.

Feel free to use letters other than x and y to denote the variables. In Example 4.1, for instance, the independent variables are t, r, and n, and the dependent variables are s, A, and C. Also, as we see in the next example, letters other than f can be used to designate functions.

Example 4.3   (a) If a function $f(x)$ is defined by the formula $f(x) = x^3 - 3x^2 + 5$, then $f(2) = 2^3 - 3 \cdot 2^2 + 5 = 1$, $f(0) = 5$, and $f(-2) = (-2)^3 - 3(-2)^2 + 5 = -15$.

(b) If a function $g(x)$ is defined by the formula $g(x) = \sqrt{x}$, then $g(1) = \sqrt{1} = 1$, $g(4) = \sqrt{4} = 2$, and $g(10) = \sqrt{10} = 3.16227766017$, approximately. In this case the only allowed values of x are those for which $x \geq 0$, because square roots of negative numbers aren't real numbers.

(c) If a function $h(x)$ is defined by the formula $h(x) = 1/(4 - x)$, then $h(1) = 1/(4 - 1) = 1/3$, $h(2) = 1/(4 - 2) = 1/2$, and $h(4) = 1/(4 - 4) = 1/0$ does not exist, because division by zero isn't permitted in algebra. Thus, $x = 4$ is the only value of x that's not allowed.

A function isn't fully known until we know precisely which real numbers are permissible values for the independent variable x. The domain is therefore an indispensable part of the concept of a function. In practice, however, most of the specific functions we deal with are defined only by formulas like the ones in Example 4.3, and nothing is said about the domain. Unless we state otherwise, the domain of such a function is understood to be the set of all real numbers x for which the formula makes sense. In part (a) of Example 4.3, this means all real numbers; in part (b), all real numbers $x \geq 0$; and in part (c), all real numbers except $x = 4$.

You're undoubtedly familiar with the idea of the *graph* (Chapter 5) of a function *f*: if we imagine the domain *D* spread out on the *x*-axis in the coordinate plane, then to each number *x* in *D* there corresponds a number $y = f(x)$, and the set of all the resulting points (x, y) in the plane is the graph. Graphs are pictures of functions that let us see these functions in their entirety, and we'll examine many in the next chapter.

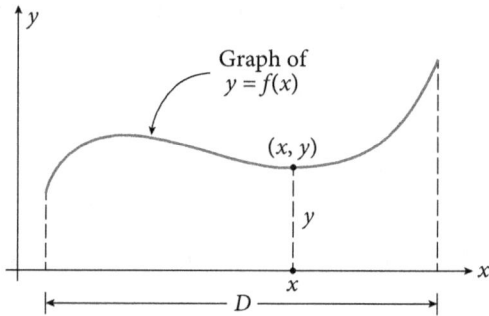

You might find it helpful to visualize a function by means of a **machine diagram**, as shown in the following figure. Here a number *x* in the domain is fed into the machine, where it's acted upon by the specific instructions built into the function *f*, and this action produces the resulting number $f(x)$. The domain is the set of all permissible inputs *x*, and the range is the set of all outputs $f(x)$.

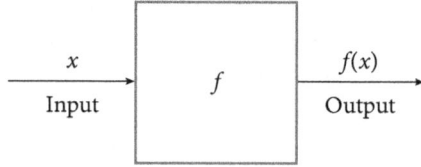

Another way to picture a function is by an **arrow diagram**, in which the domain is thought of as a certain set of points and the range as another set of points, as shown in the following figure. The arrow shows that *x* has $f(x)$ corresponding to it, and the function *f* is the complete collection of all these correspondences thought of as a mapping of the first set onto the second.

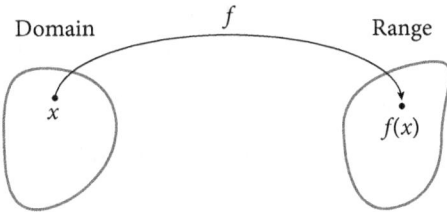

The primary tool for visualizing a function is always a graph. Machine diagrams and arrow diagrams are useful *only* if you're having trouble grasping the concept of a function.

Originally, the only functions that mathematicians considered were those defined by formulas. This restriction led to the useful intuitive idea that a function f "does something" to each number x in its domain to produce the corresponding number $y = f(x)$. Thus, if

$$y = f(x) = (x^3 + 4)^2,$$

then y is the result of applying certain specific operations to x: cube it, add 4, and then square the sum. On the other hand, the following equation is also a perfectly legitimate function which is defined by a verbal prescription instead of a formula:

$$y = f(x) = \begin{cases} 1 & \text{if } x \text{ is a rational number,} \\ 0 & \text{if } x \text{ is an irrational number.} \end{cases}$$

All that's really needed of a function is that y be determined uniquely—in any manner whatsoever—when x is specified. Beyond this, nothing is required of the rule f. In discussions that focus on ideas instead of specific functions, such generality is often an advantage. For example, in calculus this broad thinking is crucial for discovering what conditions must be imposed on an arbitrary function to guarantee that its integral exists.

Strictly speaking, the word "function" refers to the rule of correspondence f that assigns a unique number $y = f(x)$ to each number x in the domain. Mathematical sticklers emphasize the distinction between the function f and its value $f(x)$ at x. After this distinction is clearly understood, however, most people tend to use the word loosely and speak of "the function $y = f(x)$", or even "the function $f(x)$".

Functions are often **composite functions** (or **compound functions**) built up out of simpler ones. Consider the two functions

$$f(x) = x^2 + 3x \qquad \text{and} \qquad g(x) = x^2 - 1.$$

The single function that results from first applying g to x and then applying f to $g(x)$ is

$$f(g(x)) = f(x^2 - 1)$$
$$= (x^2 - 1)^2 + 3(x^2 - 1)$$
$$= x^4 + x^2 - 2.$$

Notice that $f(x^2 - 1)$ is obtained by replacing x by the entire quantity $x^2 - 1$ in the formula $f(x) = x^2 + 3x$. The symbol $f(g(x))$ is read "f of g of x" and is called a **function of a function**. If we apply the functions in the other order (first f, then g, or "g of f of x"), we have

$$g(f(x)) = g(x^2 + 3x)$$
$$= (x^2 + 3x)^2 - 1$$
$$= x^4 + 6x^3 + 9x^2 - 1,$$

so $f(g(x))$ and $g(f(x))$ are different. In special cases it can happen that $f(g(x))$ and $g(f(x))$ are the same function of x. For example, if $f(x) = 2x - 3$ and $g(x) = -x + 6$, then:

$$f(g(x)) = f(-x + 6) = 2(-x + 6) - 3 = -2x + 9,$$
$$g(f(x)) = g(2x - 3) = -(2x - 3) + 6 = -2x + 9.$$

In each of these examples two given functions are combined into a single composite function. In most practical work we proceed in the other direction, and decompose composite functions into their simpler constituents. For example, if

$$y = (x^3 + 1)^7,$$

then we can introduce an auxiliary variable u by writing $u = x^3 + 1$ and decompose the above function into the two simpler functions

$$y = u^7 \qquad \text{and} \qquad u = x^3 + 1.$$

Decompositions of this type are often useful in calculus.

In practice, functions often arise from algebraic relations between variables. Thus, an equation involving x and y determines y as a function of x if the equation is equivalent to one that expresses y *uniquely* in terms of x. For example, the equation $4x + 2y = 6$ can be solved for y, $y = 3 - 2x$, and this second equation defines y as a function of x. However, in some cases it happens that the process of solving for y leads to more than one value of y. If the equation is $y^2 = x$, for example, then we get $y = \pm\sqrt{x}$. Because this result gives two values of y for each positive value of x, the equation $y^2 = x$ doesn't by itself determine y as a function of x. If we like, we can split the formula $y = \pm\sqrt{x}$ into two separate formulas, $y = \sqrt{x}$ and $y = -\sqrt{x}$. Each of these formulas defines y as a function of x, so that out of one equation we get two functions.

The number of distinct individual functions is clearly unlimited. Most of those appearing in this book are relatively simple, however, and can be classified into a few convenient categories. The remainder of this chapter give a rough description of these categories in order of increasing complexity.

Polynomials

The simplest functions are the powers of x with nonnegative integer exponents,

$$1, x, x^2, x^3, \ldots, x^n, \ldots.$$

If a finite number of these are multiplied by constants and the results are added, then we get a general polynomial,

$$p(x) = a_0 + a_1 x + a_2 x^2 + a_3 x^3 + \cdots + a_n x^n.$$

The **degree** of a polynomial is the largest exponent that occurs in it; if $a_n \neq 0$, then the degree of $p(x)$ is n. The following examples are polynomials of degrees 1, 2, and 3:

$$y = 3x - 2, \qquad y = 1 - 2x + x^2, \qquad y = x - x^3.$$

Polynomials can be multiplied by constants, added, subtracted, and multiplied together, and the results are again polynomials.

Rational Functions

If division is also allowed, then we pass beyond the polynomials into the more inclusive class of rational functions, such as

$$\frac{x}{x^2+1}, \quad \frac{x+2}{x-2}, \quad \frac{x^3-4x^2+x+6}{x^2+x+1}, \quad x+\frac{1}{x}.$$

The general rational function is a quotient of polynomials,

$$\frac{a_0 + a_1 x + a_2 x^2 + \cdots + a_n x^n}{b_0 + b_1 x + b_2 x^2 + \cdots + b_m x^m},$$

and a specific function is rational if it is (or can be expressed as) such a quotient. If the denominator here is a nonzero constant, then this quotient is itself a polynomial. Thus, the polynomials are included among the rational functions.

Algebraic Functions

If root extractions are also allowed, then we pass beyond the rational functions into the larger class of algebraic functions. Some simple examples are

$$y=\sqrt{x}, \quad y=x+\sqrt[3]{x^2+1}, \quad y=\frac{1}{\sqrt{1-x}}, \quad y=\sqrt[4]{\frac{x+1}{x-1}}.$$

If we replace the root symbols by fractional exponents in accordance with the rules of algebra, then these functions can be written

$$y=x^{1/2}, \quad y=x+(x^2+1)^{1/3}, \quad y=(1-x)^{-1/2}, \quad y=\left(\frac{x+1}{x-1}\right)^{1/4}.$$

Transcendental Functions

Any function that's not algebraic is called **transcendental**. The transcendental functions studied in calculus are the trigonometric, inverse trigonometric, exponential, and logarithm functions. Some simple examples are

$$y=x^\pi, \quad y=a^x, \quad y=x^x, \quad y=x^{\frac{1}{x}}, \quad y=\log_a x, \quad y=\sin x.$$

Geometric Formulas

This section provides a brief review of some important functions arising in geometry. A ready grasp of the geometric formulas given in the accompanying figures is essential for coping with many examples and problems in calculus. These formulas—for the circumference and area of a circle, the total surface area and volume of a sphere, and the lateral surface area and volume of a cylinder and a cone—should be understood if possible, but remembered in any event (details can be found in Tim Hill, *Essential Geometry: A Self-Teaching Guide*).

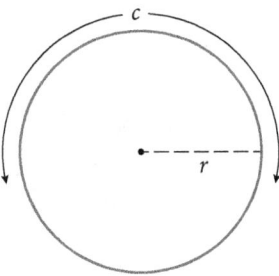

Circle
circumference $c = 2\pi r$
area $A = \pi r^2$

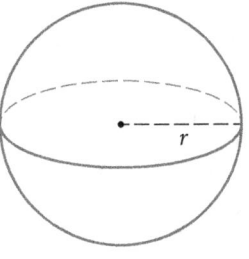

Sphere
surface area $A = 4\pi r^2$
volume $V = \frac{4}{3}\pi r^3$

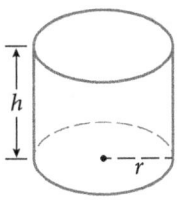

Cylinder
lateral area $A = 2\pi rh$
volume $V = \pi r^2 h$

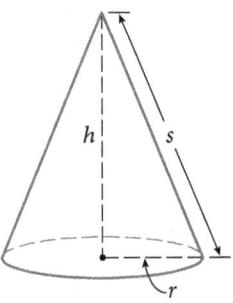

Cone
lateral area $A = \pi rs$
volume $V = \frac{1}{3}\pi r^2 h$

Each of the first four formulas, those for the circle and the sphere, defines a function of the independent variable r, in which a given positive value of r determines the corresponding value of the dependent variable.

Our attention in this book is directed at functions of a single independent variable, as previously defined and discussed. Nevertheless, we point out that each of the last four formulas, those for the cylinder and the cone, defines a function of the two variables r and h; these variables are called **independent** (of each other) because the value assigned to either needn't be related to the value assigned to the other. In special circumstances a function of this type can be expressed as a function of one variable alone. For example, if the height of a cone is known to be twice the radius of its base so that $h = 2r$, then the formula for its volume can be written as a function of r or as a function of h:

$$V = \tfrac{1}{3}\pi r^2 (2r) = \tfrac{2}{3}\pi r^3 \quad \text{or} \quad V = \tfrac{1}{3}\pi \left(\frac{h}{2}\right)^2 h = \tfrac{1}{12}\pi h^3.$$

The formulas in the preceding geometric figures also illustrate the custom of choosing letters for variables that suggest the quantities of interest, such as A for area, V for volume, r for radius, and h for height.

Problems

1. If $f(x) = 5x^2 - 3$, find:
   (a) $f(-3)$
   (b) $f(2)$
   (c) $f(0)$
   (d) $f(-\sqrt{7})$
   (e) $f(a + 3)$
   (f) $f(5t)$

2. If $g(x) = (x - 1)/(x + 1)$, find:
   (a) $g(3)$
   (b) $g(-3)$
   (c) $g(1/3)$
   (d) $g(1/a)$
   (e) $g(a + 1)$
   (f) $g(t - 1)$

3. Compute and simplify the quantity $\dfrac{f(x+h)-f(x)}{h}$ for:

   (a) $f(x) = 5x - 3$
   (b) $f(x) = x^2$
   (c) $f(x) = 1/x$

4. If $f(x) = x^3 - 3x^2 + 4x - 2$, then compute $f(1), f(2), f(3), f(0), f(-1)$, and $f(-2)$.

5. If $f(x) = 2^x$, then compute $f(1), f(3), f(5), f(0)$, and $f(-2)$.

6. If $f(x) = 4x - 3$, show that $f(2x) = 2f(x) + 3$.

7. What are the domains of $f(x) = 1/(x - 8)$ and $g(x) = x^3$? What is $h(x) = f(g(x))$? What is the domain of $h(x)$?

8. Find the domain of each of the following functions:

   (a) $\sqrt{x}$

   (b) $\sqrt{-x}$

   (c) $\sqrt{x^2}$

   (d) $\sqrt{x^2 - 4}$

   (e) $\dfrac{1}{x^2 - 4}$

   (f) $\dfrac{1}{x^2 + 4}$

   (g) $\sqrt{(x-1)(x+2)}$

   (h) $\dfrac{1}{\sqrt{(x-1)(x+2)}}$

   (i) $\sqrt{3 - 2x - x^2}$

   (j) $\sqrt{\dfrac{x}{x-2}}$

9. If $f(x) = x/(x - 1)$, compute $f(0), f(1), f(2), f(3)$, and $f(f(3))$. Show that $f(f(x)) = x$.

10. If $f(x) = 1/(1 - x)$, compute $f(0), f(1), f(2), f(f(2))$, and $f(f(f(2)))$. Show that $f(f(f(x))) = x$.

11. If $f(x) = 2^x$, use functional notation to express the fact that $2^a \cdot 2^b = 2^{a+b}$.

12. A **linear** function is one that has the form $f(x) = ax + b$, where a and b are constants. If $g(x) = cx + d$ is also linear, then is it always true that $f(g(x)) = g(f(x))$?

13. A **quadratic** function is one that has the form $f(x) = ax^2 + bx + c$, where a, b, c are constants and $a \neq 0$.

    (a) Find the values of the coefficients a, b, c if $f(0) = 3, f(1) = 2, f(2) = 9$.

    (b) Show that, no matter what values are given to the coefficients, a, b, c, the range of a quadratic function can't be the set of all real numbers.

14. Split the equation $2x^2 + 2xy + y^2 = 3$ into two equations, each of which determines y as a function of x.

15. The equal sides of an isosceles triangle are 2. If x is the base, express the area A as a function of x.

16. A rectangle whose base has length x is inscribed in a fixed circle of radius a. Express the area A of the rectangle as a function of x.

17. Express the area A of a square as a function of the length of one side s and as a function of the perimeter p.

18. Express the area A of a circle as a function of its circumference c.

19. Express the height h of an equilateral triangle as a function of its base b.

20. A cylinder has fixed volume V. Express its total surface area as a function of the radius r of its base.

21. A farmer has 100 feet of fencing with which to build a rectangular chicken pen. If x is the length of one side of the pen, show that the enclosed area is

$$A = 50x - x^2 = 625 - (x - 25)^2.$$

Use this result to find the largest possible area and the lengths of the sides that yield this largest area.

22. Find the domain of each of the following functions:

(a) $5 - x$

(b) $\dfrac{x}{2x - 3}$

(c) $\sqrt{3x - 2}$

(d) $\sqrt{5 - 3x}$

(e) $\dfrac{x + 7}{x^2 - 9}$

(f) $\sqrt[3]{x}$

(g) $\sqrt{9 - 4x^2}$

(h) $\dfrac{1}{\sqrt{x + 3}}$

(i) $\sqrt{7x^2 + 5}$

23. If $f(x) = ax + b$, show that $f\left(\dfrac{x_1 + x_2}{2}\right) = \dfrac{f(x_1) + f(x_2)}{2}$.

24. If $f(x) = \sqrt[3]{x}$, then what function $g(x)$ has the property that $g(f(x)) = x$?

25. A cylinder has fixed total surface area A. Express its volume as a function of the radius r of its base.

5   Graphs

In the preceding chapter we learned that the concept of a function can be summarized as follows. If x and y are two variables which are related in such a way that whenever a suitable numerical value is assigned to x there is determined a single corresponding numerical value for y, then y is called a *function of x* and this is expressed by writing $y = f(x)$. The letter f symbolizes the function itself, which is the operation or rule of correspondence that yields y when applied to x. For practical reasons, however, we prefer to speak of the "function $y = f(x)$" rather than "the function f". A function isn't a formula and needn't be specified by a formula—even though most are.

Now for graphs. The well-known proverb "One picture is worth a thousand words" expresses a basic truth about the study of mathematics. Cultivate the habit of thinking graphically as second nature. In the study of functions, this means drawing graphs.

It's often possible to think concretely of the graph of a function $y = f(x)$ as the path of a moving point, as illustrated in the following figure.

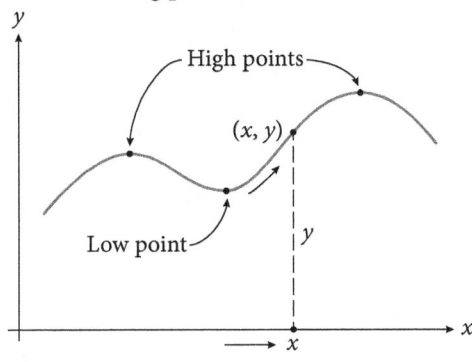

The independent variable x can be seen as a point moving along the x-axis from left to right; each x determines a value of the dependent variable y, which is the height of the point (x, y) above the x-axis. The graph of the function is simply the path of the point (x, y) as it moves across the coordinate plane, sometimes rising and sometimes falling, and in general varying in height according to the dictates of the function. The graph as a whole is intended to provide a clear overall picture of this variation. The graph shown above happens to be a smooth curve with two high points and one low point, but many diverse phenomena are possible.

In this chapter, we'll discuss the graphs of a few representative examples of the types of functions described in the preceding chapter.

Polynomials

We've seen that the simplest polynomials are the powers of x with non-negative integral exponents,

$$1, x, x^2, x^3, \ldots, x^n, \ldots .$$

As we know, the graph of $y = 1$ is the horizontal straight line through the point $(0, 1)$, and the graph of $y = x$ is the straight line through the origin with slope 1.

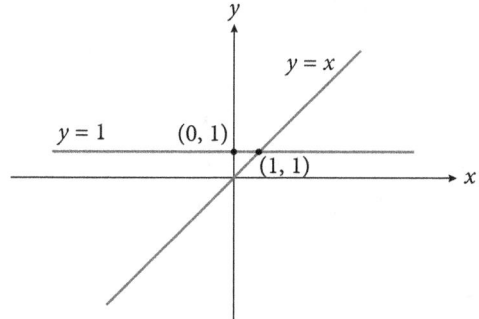

For larger values of the exponent n, the graphs of $y = x^n$ are of two distinct types, depending on whether n is even or odd:

$$y = x^2, x^4, x^6, \ldots$$

and

$$y = x^3, x^5, x^7, \ldots.$$

These two types are shown in the following figure. As n increases, these curves become flatter near the origin and steeper outside the interval $[-1, 1]$.

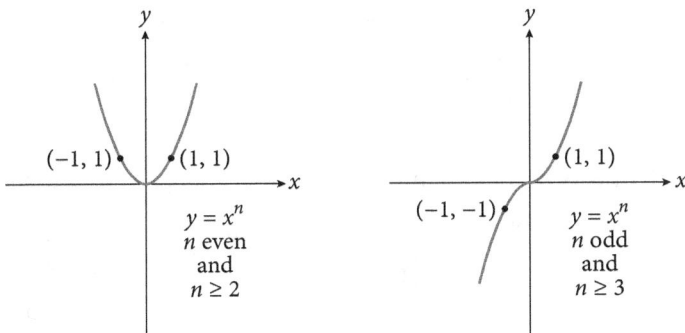

We already know that the graphs of all first- and second-degree polynomials, such as

$$y = 2x - 1$$

and

$$y = 3x^2 - 2x + 1,$$

are straight lines and parabolas. These graphs are easy to draw—without plotting points—by using the ideas in Chapters 2 and 3.

Next, we need some new terminology. A **zero** of a function $y = f(x)$ is a root of the corresponding equation $f(x) = 0$. Geometrically, the zeros of this function (if it has any) are the values of x at which its graph crosses or touches the x-axis; they are the x-intercepts of this graph.

Now consider the general second-degree polynomial

$$y = ax^2 + bx + c, \quad a \neq 0. \quad (1)$$

As we know, the graph of this function is a parabola for all values of the coefficients. If we assume that $a > 0$, so that the parabola opens upward, then three possibilities exist for the zeros of (1), as shown in the following figure.

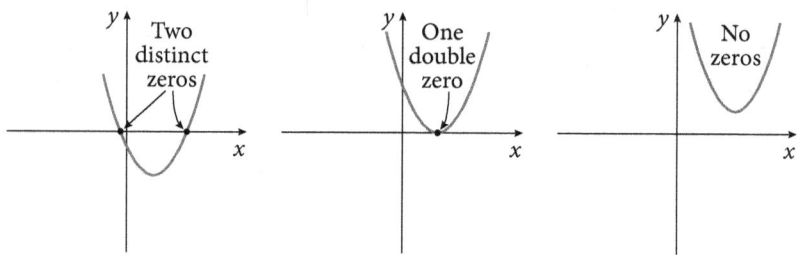

Because the roots of the quadratic equation $ax^2 + bx + c = 0$ are given by the quadratic formula

$$x = \frac{-b \pm \sqrt{b^2 - 4ac}}{2a},$$

it's clear that the three possibilities in the figure above correspond to the algebraic conditions $b^2 - 4ac > 0$, $b^2 - 4ac = 0$, $b^2 - 4ac < 0$.

Graphing polynomials of degree $n \geq 3$ isn't easy. The following example suggests some useful ideas.

Example 5.1   The graph of

$$y = x^3 - 3x \quad (2)$$

is shown in the following figure.

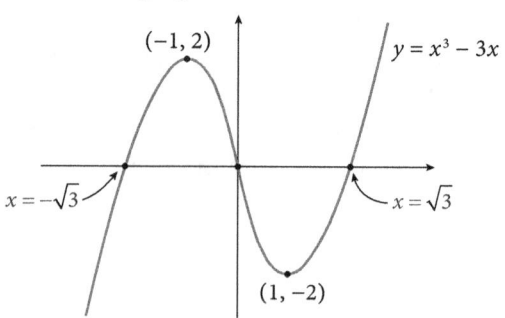

At present we have no methods available for discovering such important features of this curve as the precise location of the indicated high and low points. These methods are part of calculus. Nevertheless, we can make a few observations that provide enough details to sketch the graph.

If we write (2) in factored form, as

$$y = x(x^2 - 3) = x(x + \sqrt{3})(x - \sqrt{3}), \qquad (3)$$

then its zeros are obviously $0, -\sqrt{3}, \sqrt{3}$. These three numbers partition the x-axis into four intervals, as shown in the following figure, and inspecting the factors of (3) tells us that in each interval y has the sign given in the figure.

Determining the sign of y correctly is crucial. The details are:

For $x < -\sqrt{3}$,
   x is negative,
   $x + \sqrt{3}$ is negative, and
   $x - \sqrt{3}$ is negative,
   so their product y is negative.

For $-\sqrt{3} < x < 0$,
   x is negative,
   $x + \sqrt{3}$ is positive, and
   $x - \sqrt{3}$ is negative,
   so their product y is positive.

For $0 < x < \sqrt{3}$,
   x is positive,
   $x + \sqrt{3}$ is positive, and
   $x - \sqrt{3}$ is negative,
   so their product y is negative.

For $x > \sqrt{3}$,
   x is positive,
   $x + \sqrt{3}$ is positive, and
   $x - \sqrt{3}$ is positive,
   so their product y is positive.

We therefore know, for each interval, whether the graph of (2) lies above or below the x-axis (see the figure of the graph above). Analyzing the problem at this level is often useful in general curve sketching.

Our second observation concerns the behavior of the graph of (2) when x is numerically large (that is, far to the right or far to the left on the x-axis). If (2) is written in the form

$$y = x^3\left(1 - \frac{3}{x^2}\right), \quad x \neq 0,$$

then for large positive or negative values of x the expression in parentheses is nearly 1, so y is close to x^3. In geometric terms, when x is large, the graph of (2) is close to the graph of $y = x^3$, as the following figure suggests. In particular, the graph of (2) rises on the far right and falls on the far left.

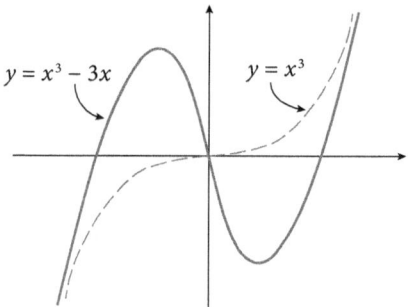

You can always sketch a graph by plotting many points laboriously and then joining them by a reasonable curve. Yet you should use this brute-force procedure only as a last resort, when more-imaginative methods fail. The important features of functions and their graphs are much more clearly revealed by the qualitative approach to curve sketching in Example 5.1.

Rational Functions

Example 5.2 The simplest rational function that isn't a polynomial is

$$y = \frac{1}{x}. \qquad (4)$$

Examine (4) and notice the following facts: y is undefined when $x = 0$; y is positive when x is positive, and is small when x is large and large when x is near 0 on the right; y is negative when x is negative, and is small when x is large and large when x is near 0 on the left. The graph

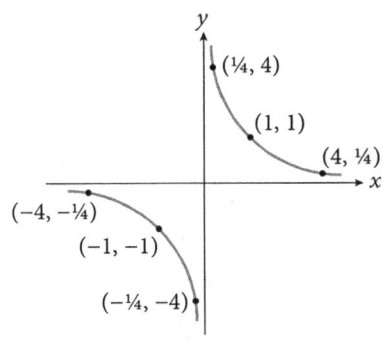

of (4) shown in the accompanying figure is a pictorial representation of these statements. In this case the graph is also easy to sketch by plotting a few points, as shown. However, you will profit much more from visualizing the behavior of such a function on the various parts of its domain and drawing what you see in your mind's eye.

A straight line is called an **asymptote** of a curve if, as a point moves out along an extremity of the curve, the distance from this point to the line approaches 0. It's clear that both the x-axis and the y-axis are asymptotes of the graph shown above. The behavior of the function (4) at and near the point $x = 0$—that is, the fact that y is undefined at $x = 0$ and "becomes infinite" near $x = 0$—is described by calling this point an **infinite discontinuity** of the function.

Example 5.3 In the case of the function

$$y = \frac{x}{x-1}, \qquad (5)$$

it's clear that the point $x = 1$ is significant, since y is undefined at $x =$

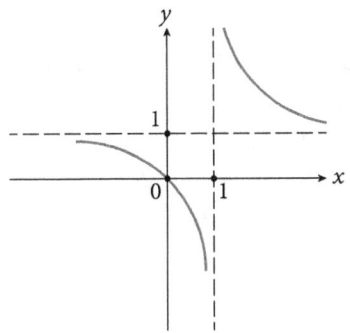

1 and is large in absolute value when x is near 1 ($x = 1$ is an infinite discontinuity). Also, y is near 1 and slightly greater than 1 when x is large and positive, and is near 1 and slightly less than 1 when x is large and negative (to see why, plot a few specific values). These observations suggest drawing the vertical and horizontal guidelines shown

in the accompanying figure. If we notice that $y = 0$ when $x = 0$, and use the method of Example 5.1 to find the sign of y in each of the intervals $-\infty < x < 0$, $0 < x < 1$, and $1 < x$, then the graph as given in the figure is easy to sketch. The lines $x = 1$ and $y = 1$ are both asymptotes.

Example 5.4   The function

$$y = \frac{x}{x^2 - 3x + 2} = \frac{x}{(x-1)(x-2)} \qquad (6)$$

is similar to (5) but more complicated. The factored form of the denominator reveals two infinite discontinuities at $x = 1$ and $x = 2$. Again, $y = 0$ when $x = 0$, but this time y is small when x is large because the degree of the denominator is greater than that of the numerator. If we combine these facts with the observable sign of y in each of the intervals $-\infty < x < 0$, $0 < x < 1$, $1 < x < 2$, and $2 < x$ (using the method of Example 5.1 for each

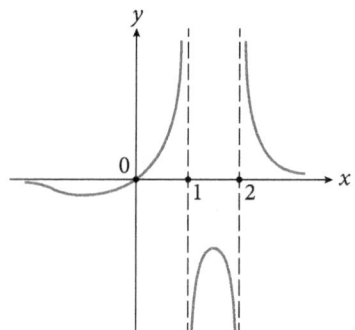

partitioned interval), then it's fairly straightforward to sketch the graph as shown in the accompanying figure. There's evidently a high point between 1 and 2, and a low point to the left of 0, but without the tools of calculus we're unable to determine the precise location of these points (in fact, they occur at $x = \sqrt{2}$ and $x = -\sqrt{2}$).

Example 5.5   The function

$$y = x + \frac{1}{x} \qquad (7)$$

has an infinite discontinuity at $x = 0$, and is positive or negative depending on whether x is positive or negative. For small positive x's, the first term on the right of (7) is negligible and the second term is large; and for large positive x's, the second term is negligible and y is approximately equal to x. We therefore sketch the part of the graph in the right half-plane as follows: draw the guideline $y = x$; insert the two extremities of the curve, approaching this guideline and the positive *y*-axis, as suggested by the behavior just stated; and connect these extremities in a reasonable way in the middle, where this part of the graph has an obvious low point. The function behaves similarly—with a corresponding high point—for negative values of *x*. The *y*-axis and the line *y* = *x* are both asymptotes. A sketch of the graph with these characteristics is shown in the accompanying figure.

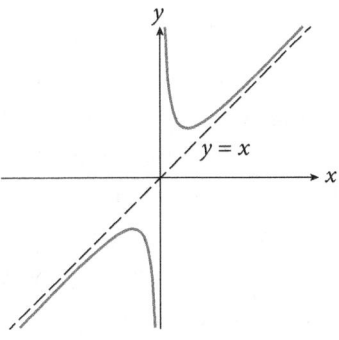

Example 5.6 The denominator of

$$y = \frac{x}{x^2 + 1} \qquad (8)$$

is positive (in fact ≥ 1) for all *x*, so *y* = 0 when *x* = 0, *y* is positive when *x* is positive, and *y* is negative when *x* is negative. Also, *y* is small when *x* is large, because the degree of the denominator is greater than that of the numerator (notice that when the numerator *x* is large, the denominator $x^2 + 1$ is huge, so *y* is small). These properties of the function force the graph to have the shape shown in the accompanying figure, with one high point and one low point.

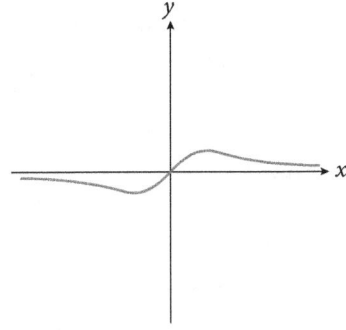

Example 5.7 In considering the function

$$y = \frac{x^2 - 1}{x - 1}, \qquad (9)$$

it's natural to factor the numerator, obtaining

$$y = \frac{(x+1)(x-1)}{x-1},$$

and then to cancel the common factor, which yields

$$y = x + 1. \qquad (10)$$

This cancellation is valid *except when x = 1*. At this point the value of (10) is 2, but (9) has no value ($y = 0/0$, which is meaningless). To graph (9), we therefore draw the straight line (10) and delete the single point (1, 2), as shown in the accompanying figure.

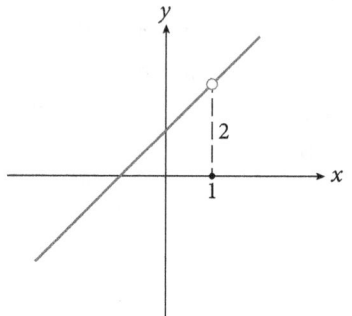

Two functions $y = f(x)$ and $y = g(x)$ are said to be **equal** if they have the same domain and if $f(x) = g(x)$ for every x in their common domain. Accordingly, the functions (9) and (10) aren't equal, because they have different domains—the point $x = 1$ is in the domain of (10) but isn't in the domain of (9). The fact that the graph of (9) has a gap (or hole) corresponding to $x = 1$ is expressed by saying that (9) is **discontinuous** at $x = 1$, or has a **discontinuity** at this point.

A note on the algebra of cancellation: to "cancel" a common *factor*, as above, is OK:

$$\frac{a\cancel{c}}{b\cancel{c}} = \frac{a}{b} \quad \text{if } c \neq 0. \quad \textit{Correct}$$

But "cancelling" a common *term* is wrong, as in

$$\frac{a+\cancel{c}}{b+\cancel{c}} = \frac{a}{b}. \quad \textit{Wrong!}$$

Otherwise, $(1 + 10)/(2 + 10)$ would equal 1/2, which it obviously does not.

Algebraic Functions

Example 5.8 The functions

$$y = \sqrt{x} \quad \text{and} \quad y = \sqrt{25 - x^2} \qquad (11)$$

can be obtained by solving the equations

$$y^2 = x \quad \text{and} \quad x^2 + y^2 = 25 \qquad (12)$$

for y and choosing the positive square roots. We know that the graphs of equations (12) are a parabola and a circle, as shown in the following figure, so the graphs of (11) are the parts of these curves that lie on or above the x-axis.

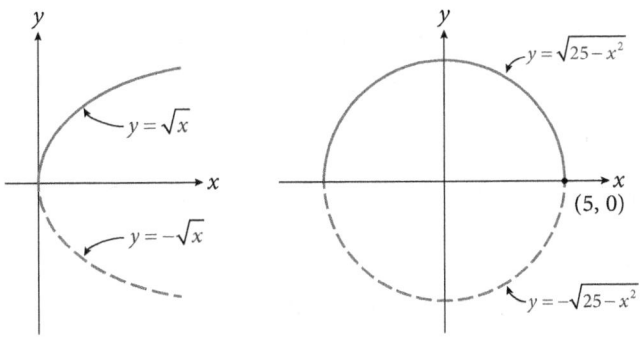

Example 5.9 The graph of the absolute value function

$$y = |x|$$

is easy to draw, as shown in the following figure. This function is algebraic because $|x| = \sqrt{x^2}$ for every value of x.

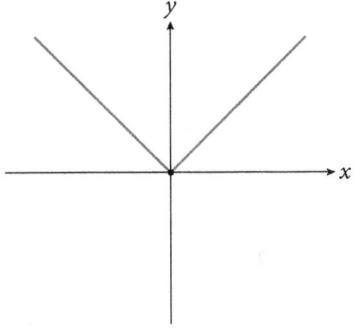

As these examples show, many of the basic features of a function are made clear by sketching its graph. We're interested less in sketches of high accuracy than in those that display broad general features: where the graph rises and falls, the presence of gaps, the presence of high points and low points, and its approximate shape. Formulas are obviously important in the study of functions—indispensable, in fact, whenever we need exact quantitative results. But never forget that the primary aim of mathematics is *insight*, and graphs are invaluable aids for gaining visual insight into the characteristics of functions.

Problems

1. Sketch the graphs of the following polynomials, paying special attention to the location of their zeros and their behavior for large values of x:

   (a) $y = x^2 + x - 2$

   (b) $y = x^3 - 3x^2 + 2x$

   (c) $y = (1 - x)(2 - x)(3 - x)$

   (d) $y = x^4 - x^2$

   (e) $y = x^4 - 5x^2 + 4$

2. Sketch the graphs of the following rational functions:

   (a) $y = \dfrac{1}{x^2}$

   (b) $y = \dfrac{1}{x^3}$

   (c) $y = x^2 + \dfrac{1}{x}$

   (d) $y = x^2 + \dfrac{1}{x^2}$

   (e) $y = \dfrac{1}{x^2 + 1}$

   (f) $y = \dfrac{x^2}{x^2 + 1}$

   (g) $y = \dfrac{1}{x^2 - 1}$

   (h) $y = \dfrac{x}{x^2 - 1}$

   (i) $y = \dfrac{x^2}{x^2 - 1}$

3. Sketch the graphs of the following algebraic functions:

(a) $y = \sqrt{(x-1)(3-x)}$

(b) $y = \dfrac{1}{\sqrt{(x-1)(3-x)}}$

(c) $y = \dfrac{1}{\sqrt{x-1}}$

(d) $y = \sqrt{\dfrac{x}{3-x}}$

(e) $y = \sqrt{\dfrac{4-x}{x-2}}$

(f) $y = \sqrt{\dfrac{x-4}{x-2}}$

4. Sketch the graphs of the following functions:

(a) $y = |x|/x$

(b) $y = |2x + 3|$

(c) $y = x + |x|$

(d) $y = 2x + |x|$

(e) $y = x - |x|$

(f) $y = 1 + x - |x|$

(g) $y = |x^2 - 1|$

5. Are any of the following pairs of functions equal?

(a) $f(x) = x/x,\ g(x) = 1$

(b) $f(x) = x^2 - 1,\ g(x) = (x + 1)(x - 1)$

(c) $f(x) = x,\ g(x) = \sqrt{x^2}$

(d) $f(x) = x,\ g(x) = (\sqrt{x})^2$

6. If n is any integer ≥ 1, then show that there exists a polynomial of degree n with n zeros. If n is even, then find a polynomial of degree n with no zeros. If n is odd, then find one with only one zero.

7. A function f is said to be **even** if $f(-x) = f(x)$ for every x in its domain, and it's said to be **odd** if $f(-x) = -f(x)$ for every x in its domain (in each case, it's understood that $-x$ is in the domain of f whenever x is). Determine whether each of the following functions is even, odd, or neither:

(a) $f(x) = x^3$

(b) $f(x) = x(x^3 + x)$

(c) $f(x) = |x|$

(d) $f(x) = x + \dfrac{1}{x}$

(e) $f(x) = x^2 + \dfrac{1}{x}$

(f) $f(x) = \dfrac{x^3 + x}{x^2 + 1}$

(g) $f(x) = x^5 + 1$

(h) $f(x) = x(x + 1)$

8. What is the distinguishing feature of the graph of an even function? Of an odd function?

9. What can be said about:
   (a) the product of two even functions?
   (b) the product of two odd functions?
   (c) the product of an even function and an odd function?

10. Write a second-degree polynomial whose values at 1, 2, and 3 are π, $\sqrt{3}$, and 550.

11. The symbol $[x]$ denotes the greatest integer that is $\leq$ a real number x. Thus, $[1] = 1$, $[2.1] = 2$, $[\pi] = 3$, and $[-1.7] = -2$. Sketch the graphs of the following functions:
   (a) $y = [x]$
   (b) $y = x - [x]$
   (c) $y = \sqrt{x - [x]}$
   (d) $y = [x] + \sqrt{x - [x]}$

6

Trigonometry

Periodic phenomena are found everywhere in the world around us—alternating currents, revolving planets, swinging pendulums, vibrating springs, and so on—and scientists describe these phenomena by using trigonometric functions. For this and other reasons, a grounding in trigonometry is required for anyone studying calculus.

This chapter introduces the fundamental ideas of trigonometry, including the radian measure of angles and the definitions and simpler properties of the important functions $\sin \theta$ and $\cos \theta$ (the Greek letter θ is pronounced "theta"). The other four trigonometric functions—$\tan \theta$, $\cot \theta$, $\sec \theta$, and $\csc \theta$—are described briefly later in this chapter (a broader look at this subject can be found in Tim Hill, *Essential Trigonometry: A Self-Teaching Guide*).

Right Triangle Trigonometry

In elementary trigonometry the sine and cosine of an acute angle θ are first defined as ratios of sides in a right triangle, as follows:

$$\sin \theta = \frac{\text{opposite side}}{\text{hypotenuse}} = \frac{a}{h},$$

$$\cos \theta = \frac{\text{adjacent side}}{\text{hypotenuse}} = \frac{b}{h}.$$

Because similar triangles have proportional sides, the values of $\sin \theta$ and $\cos \theta$ depend on only the size of the acute angle θ, and not at all on the size of the right triangle itself.

Example 6.1 We know from geometry that in a 30°–60° right triangle, the side opposite the 30° angle is half the hypotenuse. This fact lets us draw the familiar right triangles shown in the following figure.

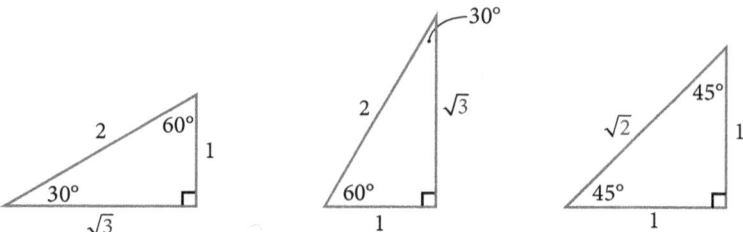

From these triangles we see that

$$\sin 30° = \frac{1}{2}, \quad \sin 60° = \frac{\sqrt{3}}{2}, \quad \sin 45° = \frac{1}{\sqrt{2}},$$

$$\cos 30° = \frac{\sqrt{3}}{2}, \quad \cos 60° = \frac{1}{2}, \quad \cos 45° = \frac{1}{\sqrt{2}}.$$

It's customary to rationalize the denominators on the right by writing

$$\frac{1}{\sqrt{2}} = \frac{1}{\sqrt{2}} \cdot \frac{\sqrt{2}}{\sqrt{2}} = \frac{1}{2}\sqrt{2}.$$

These ideas anchor **right triangle trigonometry**, in which angles are measured in degrees and sines and cosines are defined only for acute angles of right triangles. In the equivalent forms

$$a = h \sin \theta \quad \text{and} \quad b = h \cos \theta,$$

these definitions have some applications in geometry and physics but for calculus the limitations of this approach are severe. We start anew with a self-contained development of **analytic trigonometry**, which frees trigonometric functions from their dependence on right triangles and defines them as real-valued functions of a real variable.

Consider the motion of an object oscillating up and down at the end of a spring. If this motion is described by the position function

$$s = f(t) = \cos t,$$

which gives the object's position s as a function of the time t, then it makes little sense to think of t as an angle and measure its values in degrees. Instead, we must consider what cos t means when t isn't an angle but a *number*—in this case the number of seconds that have elapsed since the motion began when $t = 0$.

Radian Measure

In elementary mathematics and everyday life, angles are measured in degrees, with 90° measuring a right angle. But the degree is an arbitrary measure inherited from ancient astronomers, and its use in calculus would make many formulas intolerably messy. Calculus uses a much more natural and convenient system called **radian measure**, which is defined in terms of how much arc an angle cuts off on a circle.

In this system the unit of angle measurement is called the **radian**. One radian is the angle which, placed at the center of a circle, **subtends** (cuts off) an arc whose length equals the radius. More generally, the number of radians θ in an arbitrary central angle is defined to be the ratio of the length s of the subtended arc to the radius r, $\theta = s/r$, so that $s = r\theta$.

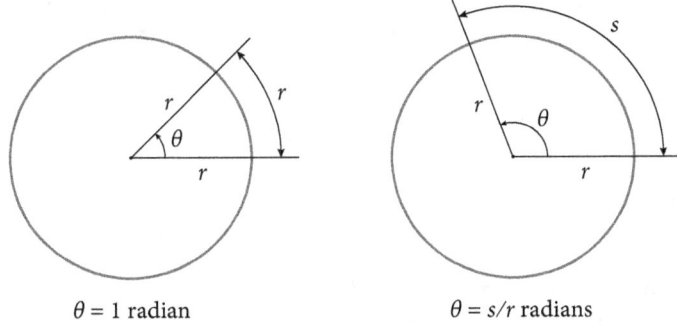

$\theta = 1$ radian $\theta = s/r$ radians

In a **unit circle** ($r = 1$), a central angle of θ radians subtends an arc of length $s = \theta$. Because the circumference of a circle is $c = 2\pi r$, a complete central angle of 360° is equivalent to $2\pi r/r = 2\pi$ radians. Thus,

$$2\pi \text{ radians} = 360° \qquad \text{or} \qquad \pi \text{ radians} = 180°;$$

and it follows from this that

$$1 \text{ radian} = \left(\frac{180}{\pi}\right)° \approx 57.296°, \quad 1° = \frac{\pi}{180} \approx 0.0175 \text{ radians}.$$

Furthermore, 90° = π/2, 60° = π/3, 45° = π/4, and 30° = π/6, where we follow the convention of omitting the word "radian" in using radian measure. It's a good idea to memorize these common conversions with the aid of the diagrams in the following figure.

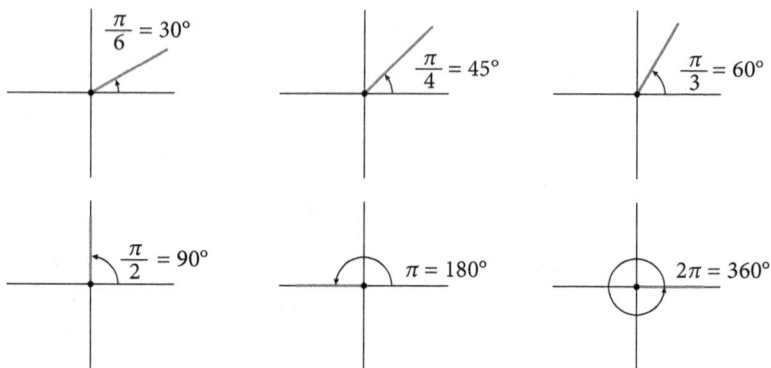

Other commonly used angles include 135° = 3π/4, 225° = 5π/4, 270° = 3π/2, and 315° = 7π/4. To gain comfort with trigonometry, you should "think in radians" rather than mentally translate from degrees to radians (the same principle applies to learning a foreign language). Some key facts are:

- The angles of a triangle sum to π radians

- A right angle is π/2 radians

- Each angle of an equilateral triangle is π/3 radians

- The line $x = y$ in the xy-plane makes an angle of π/4 radians with the positive x-axis

- In a right triangle with a hypotenuse of length 1 and another side of length ½, the angle opposite the side with length ½ is π/6 radians

- One complete rotation around a circle is 2π radians

Definitions of Sine and Cosine

We approach trigonometry by way of analytic geometry. Consider the unit circle $x^2 + y^2 = 1$ in the xy-plane and let θ be an arbitrary real number.

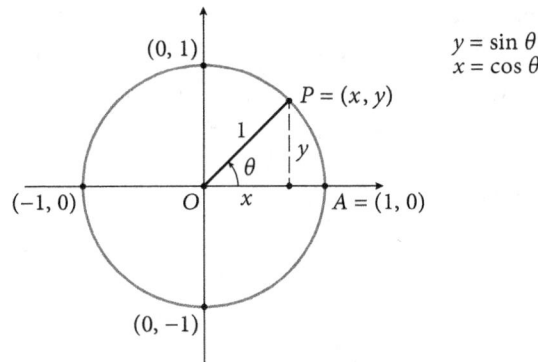

If θ is positive, then let the radius OP start in the position OA and revolve counterclockwise through θ radians. Thus, $\theta = \pi$ produces half a revolution and $\theta = 2\pi$ produces a complete revolution, both counterclockwise. If θ is negative, then we form the positive number $-\theta$ and let OP revolve clockwise through $-\theta$ radians.

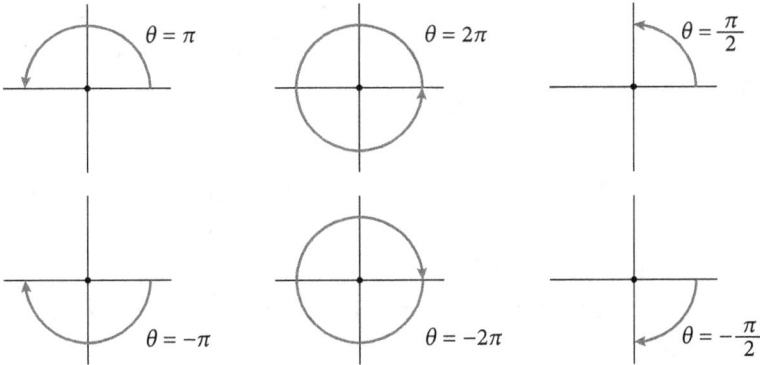

In this way, each real number θ (positive, negative, or zero) determines a unique position of the radius OP, and therefore a unique point $P = (x, y)$ with the property that $x^2 + y^2 = 1$.

The sine and cosine of θ are now defined to be

$$\sin \theta = y \qquad \text{and} \qquad \cos \theta = x.$$

The word "sine", *sinus* in Latin, is a corruption of an Arabic word meaning "chord" or "bowstring". Because sin and cos are the names of functions, the proper notation should be sin(θ) and cos(θ), just as we write $f(\theta)$ when the function is f. In the case of trigonometric functions, however, it's customary to omit the parentheses. It's evident from the definition that $-1 \leq \sin\theta \leq 1$, and similarly for $\cos\theta$. The algebraic signs of these quantities depend on which quadrant of the plane the point P lies in, as shown in the following figure.

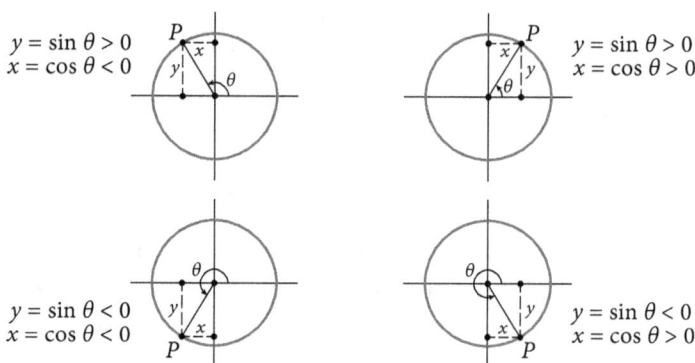

For values of θ such that $0 < \theta < \pi/2$, these definitions agree with the right triangle definitions given earlier, because in the right triangle we have $\sin\theta = y = y/1 = $ (opposite side)/(hypotenuse) and $\cos\theta = x = x/1 = $ (adjacent side)/(hypotenuse).

Identities

The trigonometric identities are the core relationships and properties of the trigonometric functions. You don't *have* to memorize any of these identities—with a little algebra, they can all be derived by using the definitions and diagrams given earlier. For the following identities, the Greek letters θ and φ (theta and phi) denote two arbitrary angles.

The angles θ and $-\theta$ are generated by the same rotation but in opposite directions, so that the endpoints of their terminal sides lie on the same vertical line. If we compare the angles θ and $-\theta$ in the following figure, then we see that

$$\sin(-\theta) = -\sin\theta \qquad \text{and} \qquad \cos(-\theta) = \cos\theta \qquad (1).$$

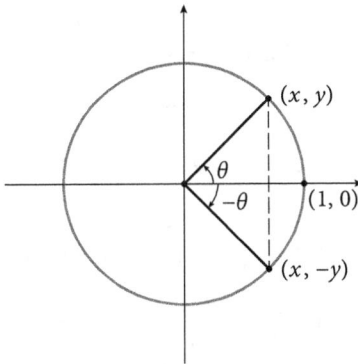

The equation $x^2 + y^2 = 1$, or equivalently $y^2 + x^2 = 1$, translates into the important identity

$$\sin^2 \theta + \cos^2 \theta = 1. \qquad (2)$$

The odd notation $\sin^2 \theta$ is the standard way of writing the square of the number $\sin \theta$—that is, $(\sin \theta)^2$—and similarly for $\cos^2 \theta$.

The **addition formulas** are

$$\sin(\theta + \varphi) = \sin \theta \cos \varphi + \cos \theta \sin \varphi, \qquad (3)$$

$$\cos(\theta + \varphi) = \cos \theta \cos \varphi - \sin \theta \sin \varphi. \qquad (4)$$

A proof of (3) for a restricted case is given below, but first the other identities.

Setting $\theta = \varphi$ in (3) and (4) yields the **double-angle formulas**

$$\sin 2\theta = 2 \sin \theta \cos \theta, \qquad (5)$$

$$\cos 2\theta = \cos^2 \theta - \sin^2 \theta. \qquad (6)$$

Finally, if we write (2) and (6) together as

$$\cos^2 \theta + \sin^2 \theta = 1,$$

$$\cos^2 \theta - \sin^2 \theta = \cos 2\theta,$$

then by adding and subtracting we get the **half-angle formulas**

$$\cos^2 \theta = \tfrac{1}{2}(1 + \cos 2\theta), \qquad (7)$$

$$\sin^2 \theta = \tfrac{1}{2}(1 - \cos 2\theta). \qquad (8)$$

The **subtraction formulas** are derived by replacing φ by $-\varphi$ in the addition formulas (3) and (4) and applying (1):

$$\sin(\theta - \varphi) = \sin \theta \cos \varphi - \cos \theta \sin \varphi,$$

$$\cos(\theta - \varphi) = \cos \theta \cos \varphi + \sin \theta \sin \varphi.$$

Now, to prove (3) for the restricted case in which θ and φ are both positive acute angles whose sum $< \pi/2$, refer to the following figure.

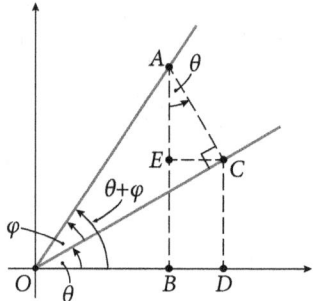

- The angles θ and φ are positive acute angles whose sum is also acute
- The arbitrary point A lies on the terminal side of $\theta + \varphi$
- AB is perpendicular to the x-axis
- AC is perpendicular to the terminal side of θ
- CD is perpendicular to the x-axis
- CE is perpendicular to AB
- The angle EAC equals θ (because both are acute angles and their sides are respectively perpendicular)

The proof of (3) is

$$\sin(\theta + \varphi) = \frac{AB}{OA} = \frac{AE + EB}{OA} = \frac{AE + CD}{OA}$$

$$= \frac{AE}{OA} + \frac{CD}{OA}$$

$$= \frac{AE}{AC} \cdot \frac{AC}{OA} + \frac{CD}{OC} \cdot \frac{OC}{OA}$$

$$= \cos\theta \sin\varphi + \sin\theta \cos\varphi.$$

A similar proof can be given for formula (4).

Values

Example 6.1 provides several first-quadrant θ's for which exact values of $\sin \theta$ and $\cos \theta$ are easy to find. These facts can also be obtained by inspecting the three parts of the following figure and remembering the Pythagorean theorem (page 8).

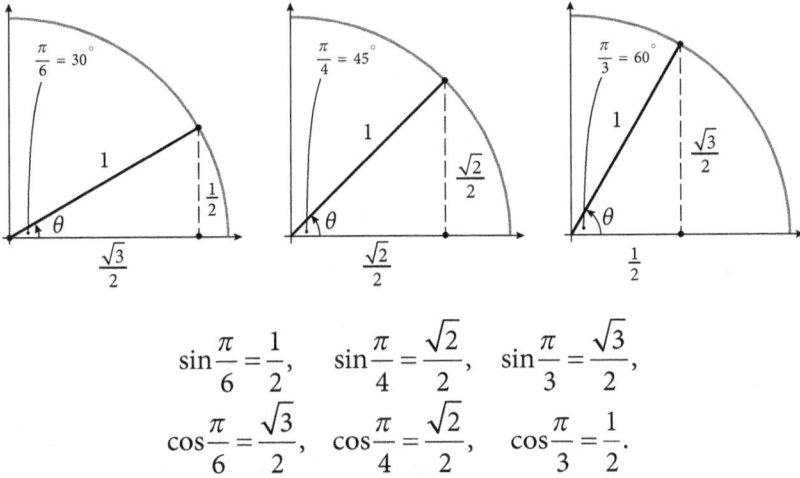

$$\sin\frac{\pi}{6} = \frac{1}{2}, \quad \sin\frac{\pi}{4} = \frac{\sqrt{2}}{2}, \quad \sin\frac{\pi}{3} = \frac{\sqrt{3}}{2},$$

$$\cos\frac{\pi}{6} = \frac{\sqrt{3}}{2}, \quad \cos\frac{\pi}{4} = \frac{\sqrt{2}}{2}, \quad \cos\frac{\pi}{3} = \frac{1}{2}.$$

Inspecting our reference figure with OP in various positions gives us similar information for the cases $\theta = 0, \pi/2, \pi, 3\pi/2, 2\pi$.

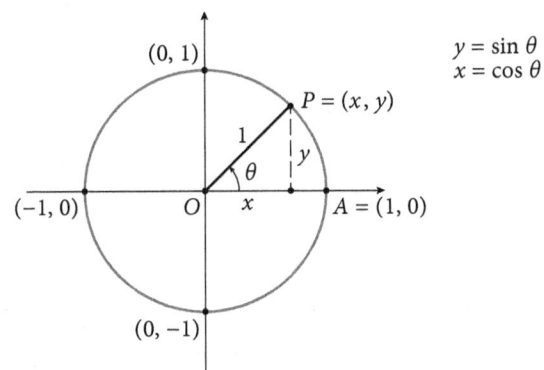

$$\sin 0 = 0, \quad \sin\frac{\pi}{2} = 1, \quad \sin \pi = 0, \quad \sin\frac{3\pi}{2} = -1, \quad \sin 2\pi = 0,$$

$$\cos 0 = 1 \quad \cos\frac{\pi}{2} = 0, \quad \cos \pi = -1, \quad \cos\frac{3\pi}{2} = 0, \quad \cos 2\pi = 1.$$

By drawing pictures and using the ideas in the preceding two figures we can now find the exact values of sin θ and cos θ for any value of θ that represents an angle one-third, one-half, or two-thirds of the way through any quadrant.

Example 6.2 The following figure indicates that $135° = 3\pi/4$ is halfway from $\pi/2$ to π.

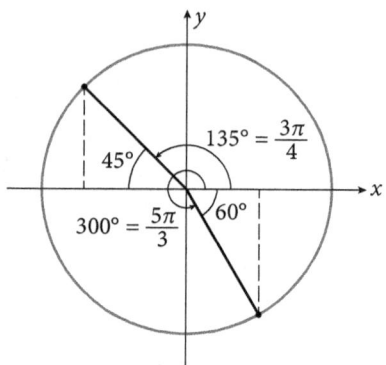

The point P is in the second quadrant with coordinates $(-\frac{1}{2}\sqrt{2}, \frac{1}{2}\sqrt{2})$, and consequently we have

$$\sin\frac{3\pi}{4} = \frac{\sqrt{2}}{2}, \quad \cos\frac{3\pi}{4} = -\frac{\sqrt{2}}{2}.$$

Similarly, $300° = 5\pi/3$ is one-third of the way from $3\pi/2$ to 2π, so P is in the fourth quadrant with coordinates $(\frac{1}{2}, -\frac{1}{2}\sqrt{3})$, and we have

$$\sin\frac{5\pi}{3} = -\frac{\sqrt{3}}{2}, \quad \cos\frac{5\pi}{3} = \frac{1}{2}.$$

Of course, most θ's are beyond the scope of these methods, and in these cases the values of sin θ and cos θ can be found by using trigonometric tables or a calculator. The problem of how these values themselves are calculated is more difficult, and require the tools of calculus.

For every θ, the numbers θ and $\theta + 2\pi$ clearly determine the same point P, so

$$\sin(\theta + 2\pi) = \sin\theta \qquad \text{and} \qquad \cos(\theta + 2\pi) = \cos\theta.$$

In other words, the values of sin θ and cos θ repeat when θ increases by 2π. We express these properties of sin θ and cos θ by saying that these functions are **periodic** with **period** 2π.

Notice that degrees are almost entirely banished from analytic trigonometry. Trigonometric *values* can be written by using degree measure or radian measure: either sin 30° or sin $\pi/6$; either cos 90° or cos $\pi/2$. But for trigonometric *functions*, such as $y = \sin \theta$ or $f(\theta) = \cos \theta$, the independent variable θ is always understood to be in radians.

Graphs

The graph of sin θ is easy to sketch by inspecting our reference figure and discerning how y varies as θ increases from 0 to 2π—that is, as the radius swings around through one complete counterclockwise revolution.

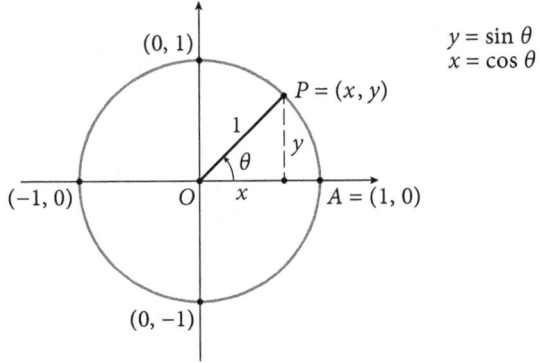

It's clear that sin θ starts at 0, increases to 1, decreases to 0, decreases further to –1, and increases to 0. This gives one complete cycle of sin θ on the interval $0 \le \theta \le 2\pi$, as shown in the following figure.

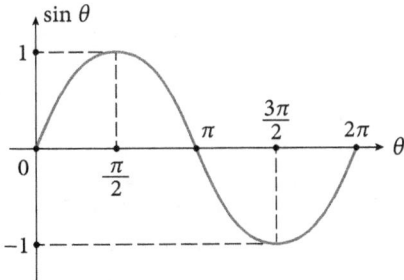

By using the periodicity of sin θ, we see that the complete graph consists of infinitely many repetitions of this cycle.

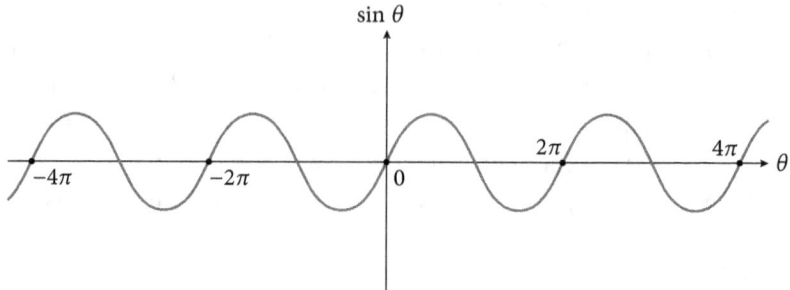

The graph of cos θ can be sketched in essentially the same way, as shown in the following figures. The main difference is that cos θ starts at 1 when $\theta = 0$, decreases to 0, decreases further to -1, increases to 0, and increases further to 1.

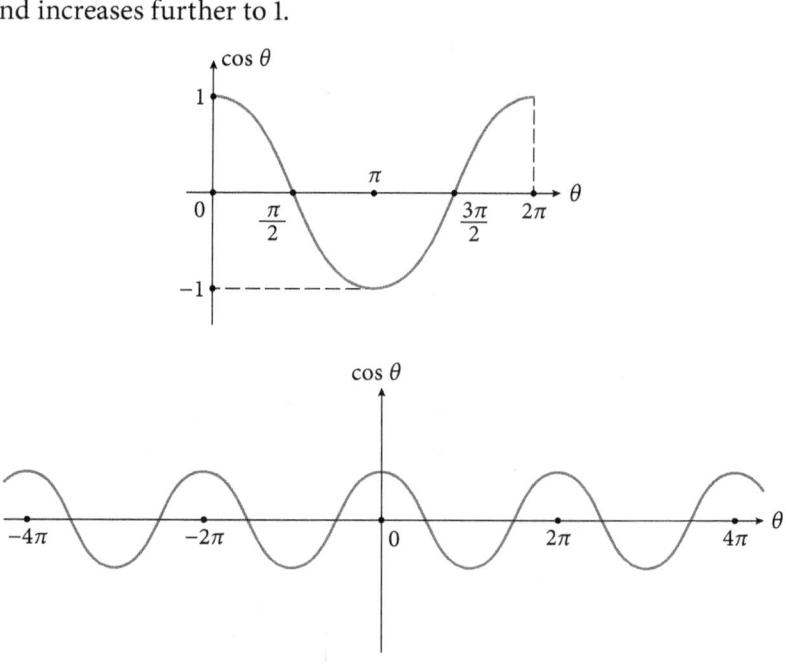

On the other hand, the graph of sin 2θ makes one complete cycle on the interval $0 \le \theta \le \pi$, because 2θ increases from 0 to 2π as θ increases from 0 to π. In other words, sin 2θ oscillates twice as fast as sin θ.

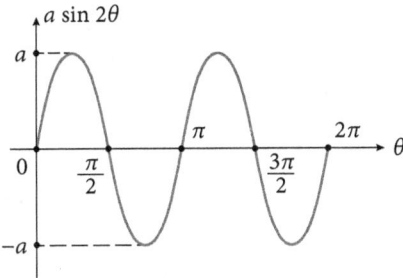

In the same way we see that sin ½θ oscillates half as fast as sin θ. In general, both sin kθ and cos kθ make one complete cycle for $0 \le k\theta \le 2\pi$, or equivalently, on the interval $0 \le \theta \le 2\pi/k$.

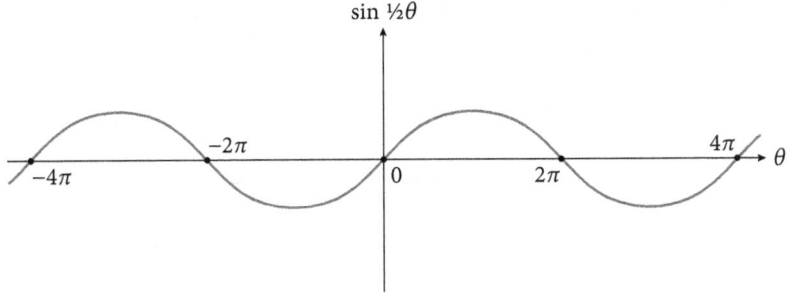

Law of Cosines

The law of cosines is a useful tool in a variety of situations in physics and geometry. It expresses the third side of a triangle in terms of two given sides a and b and the included angle θ:

$$c^2 = a^2 + b^2 - 2ab \cos \theta.$$

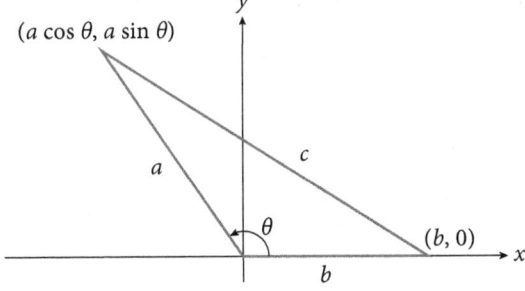

To prove this law, place the triangle in the xy-plane as shown above and apply the distance formula (page 8) to the vertices $(a\cos\theta, a\sin\theta)$ and $(b, 0)$. The square of the side c is

$$c^2 = (a\cos\theta - b)^2 + (a\sin\theta - 0)^2$$
$$= a^2(\cos^2\theta + \sin^2\theta) + b^2 - 2ab\cos\theta$$
$$= a^2 + b^2 - 2ab\cos\theta.$$

Other Trigonometric Functions

The functions $\sin\theta$ and $\cos\theta$ are the basic trigonometric functions, but there are four others that are also important though less fundamental: the tangent, cotangent, secant, and cosecant. These functions are all defined in terms of $\sin\theta$ and $\cos\theta$:

$$\tan\theta = \frac{\sin\theta}{\cos\theta},$$
$$\cot\theta = \frac{\cos\theta}{\sin\theta},$$
$$\sec\theta = \frac{1}{\cos\theta},$$
$$\csc\theta = \frac{1}{\sin\theta}.$$

Of these functions, the most useful is $\tan\theta$. In right triangle trigonometry (page 69), the tangent of an acute angle θ is defined as follows:

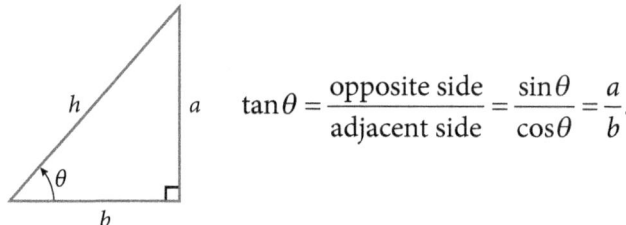

$$\tan\theta = \frac{\text{opposite side}}{\text{adjacent side}} = \frac{\sin\theta}{\cos\theta} = \frac{a}{b}.$$

The addition formula is

$$\tan(\theta + \varphi) = \frac{\tan\theta + \tan\varphi}{1 - \tan\theta\tan\varphi}.$$

To prove this fact, express $\tan(\theta + \varphi)$ in terms of $\sin(\theta + \varphi)$ and $\cos(\theta + \varphi)$.

$$\tan(\theta + \varphi) = \frac{\sin(\theta + \varphi)}{\cos(\theta + \varphi)}$$

$$= \frac{\sin\theta\cos\varphi + \cos\theta\sin\varphi}{\cos\theta\cos\varphi - \sin\theta\sin\varphi}$$

Divide both numerator and denominator by $\cos\theta\cos\varphi$.

$$= \frac{\dfrac{\sin\theta}{\cos\theta} + \dfrac{\sin\varphi}{\cos\varphi}}{1 - \left(\dfrac{\sin\theta}{\cos\theta}\right)\left(\dfrac{\sin\varphi}{\cos\varphi}\right)}$$

$$= \frac{\tan\theta + \tan\varphi}{1 - \tan\theta\tan\varphi}.$$

The graph of $\tan\theta$ differs considerably from the graphs of $\sin\theta$ and $\cos\theta$. Notice that $\tan\theta$ is periodic with period π:

$$\tan(\theta + \pi) = \frac{\sin(\theta + \pi)}{\cos(\theta + \pi)} = \frac{-\sin\theta}{-\cos\theta} = \tan\theta.$$

Inspecting our reference figure, we can get the full range of values of $\tan\theta$ by visualizing the ratio y/x as θ increases from $-\pi/2$ to $\pi/2$.

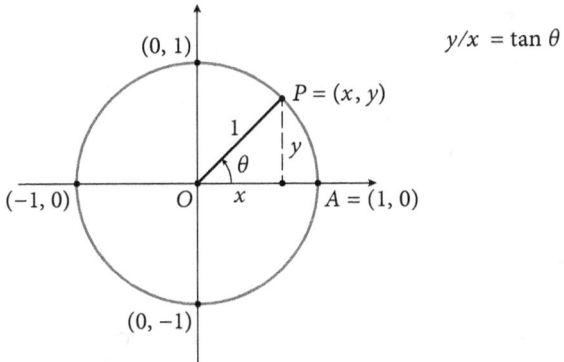

The result is the central curve shown in the following figure, and the complete graph of tan θ consists of infinitely many repetitions of this curve to the right and to the left. The fact that tan $\theta \to \infty$ as $\theta \to \pi/2$ (from the left) is often informally expressed as tan $\pi/2 = \infty$.

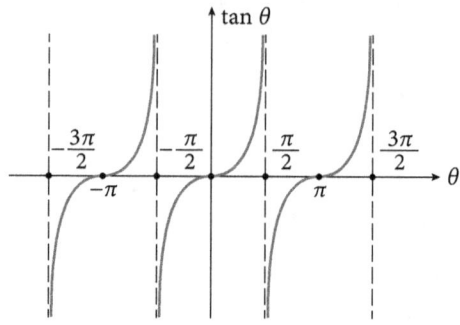

Problems

1. Convert the given angles from degrees to radians:
    (a) 15°
    (b) 150°
    (c) 1500°
    (d) –36°
    (e) 7°
    (f) –110°

2. Convert the given angles from radians to degrees:
    (a) $\pi/15$
    (b) $\pi/45$
    (c) $-\pi/36$
    (d) –3
    (e) π^2
    (f) 30

3. Find the value of the given expression without using tables or a calculator:
    (a) cos –120°
    (b) sin 780°

(c) $\sin 17\pi/3$

(d) $\cos -15\pi/4$

(e) $\sin 19\pi/6$

(f) $\cos 99\pi/4$

4. Is the given number positive, negative, or zero?

   (a) $\sin 500\pi$

   (b) $\cos 7$

   (c) $\sin 901°$

   (d) $\cos 2^4$

5. Find a formula for $\sin 3\theta$ in terms of $\sin \theta$. (Hint: $3\theta = 2\theta + \theta$.)

6. Find a formula for $\cos 3\theta$ in terms of $\cos \theta$.

7. Express each trigonometric function as a corresponding function of an angle in the first quadrant ($0 \le \theta \le \pi/2$) preceded by a + or − sign:

   (a) $\sin\left(\dfrac{9\pi}{2}\right)$

   (b) $\sin 7\pi$

   (c) $\sin\left(-\dfrac{7\pi}{3}\right)$

   (d) $\sin\left(-\dfrac{8\pi}{3}\right)$

   (e) $\cos 10\pi$

   (f) $\cos\left(\dfrac{9\pi}{4}\right)$

   (g) $\cos\left(-\dfrac{6\pi}{5}\right)$

   (h) $\sin\left(-\dfrac{11\pi}{2}\right)$

   (i) $\cos\left(\dfrac{11\pi}{3}\right)$

8. By examining the following figure, verify the identities

$$\sin(\pi - \theta) = \sin \theta$$

and

$$\cos(\pi - \theta) = -\cos \theta.$$

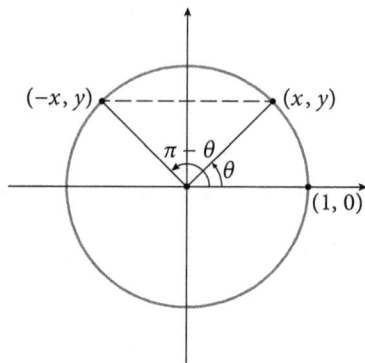

9. By examining the following figure, verify the identities

$$\sin(\pi/2 - \theta) = \cos \theta$$

and

$$\cos(\pi/2 - \theta) = \sin \theta.$$

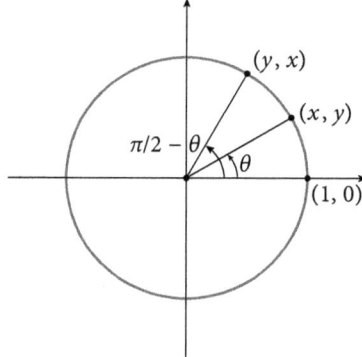

10. The half-angle formulas (7) and (8) are so called because if we set $2\theta = \alpha$, then they can be written as

$$\sin \tfrac{1}{2}\alpha = \pm\sqrt{\frac{1-\cos\alpha}{2}} \quad \text{and} \quad \cos \tfrac{1}{2}\alpha = \pm\sqrt{\frac{1+\cos\alpha}{2}}.$$

Use these formulas to find the values of $\sin 15°$ and $\cos 15°$.

11. Apply the half-angle formula for the cosine to find:
    (a) $\cos \pi/4$
    (b) $\cos 3\pi/4$

12. Use the addition formula to find $\cos 5\pi/4$ by using the fact that $5\pi/4 = \pi + \pi/4$.

13. Verify the identity for $\sin(\theta + \varphi)$ when $\theta = \pi/6$ and $\varphi = \pi/3$.

14. Find $\sin 5\pi/12$ by using the fact that $5\pi/12 = \pi/4 + \pi/6$.

15. Prove the addition formula (4) for cosine by the same method and figure used to prove addition formula (3) for sine earlier in this chapter.

16. Given a triangle with angles A, B, C and sides a, b, c opposite these angles, prove the **law of sines**:

$$\frac{\sin A}{a} = \frac{\sin B}{b} = \frac{\sin C}{c}.$$

17. What is the area of an isosceles right triangle whose hypotenuse has length h?

7

Solutions

Chapter 1

1. (a) $-2/3$, rational.
   (b) 0, integer, rational.
   (c) $45/9$, rational.
   (d) 0.75, rational.
   (e) $-\sqrt{49} = -7$, integer, rational.
   (f) $1/\pi$, irrational.
   (g) $9.000\ldots$, integer, rational.
   (h) $3^{1/2} = \sqrt{3}$, irrational.
   (i) $-20/7$, rational.
   (j) $94/7$, rational.

2. (a) Because n is even, we can write $n = 2k$ for some integer k. Then $n^2 = 4k^2 = 2(2k^2)$. Therefore n^2 is divisible by 2 and so n^2 is even by the definition of an even integer.

   (b) Because n is odd, we can write $n = 2k + 1$ for some integer k. Then $n^2 = 4k^2 + 4k + 1 = 2(2k^2 + 2k) + 1 = 2m + 1$ where $m = 2k^2 + 2k$ is an integer. Therefore n^2 is odd.

3. (a) $|7 - 18| = |-11| = 11$.
   (b) $|7| - |-18| = 7 - 18 = -11$.
   (c) Because $\pi - 3 > 0$, $|\pi - 3| = \pi - 3$.
   (d) Because $3 - \pi < 0$, $|3 - \pi| = -(3 - \pi) = \pi - 3$.
   (e) Because $x - 5 < 0$, $|x - 5| = -(x - 5) = 5 - x$.
   (f) Because $x - 5 > 0$, $|x - 5| = x - 5$.
   (g) Because $x^2 \geq 0$, $x^2 + 10 \geq 10 > 0$, so $|x^2 + 10| = x^2 + 10$.

(h) Because $x \geq 1$, we have $3x^2 \geq 3$, $1 - 3x^2 \leq 1 - 3 = -2 < 0$, so $|1 - 3x^2| = -(1 - 3x^2) = 3x^2 - 1$.

(i) Because $x < 0$, $|x| = -x$. Thus, $|x|/x = -x/x = -1$.

(j) Because $x > 0$, $|x| = x$. Thus, $|x|/x = x/x = 1$.

4. (a) $x(x - 1) > 0$ if and only if either $(x > 0$ and $x - 1 > 0)$ or $(x < 0$ and $x - 1 < 0)$ if and only if either $x > 1$ or $x < 0$.

(b) $(x - 1)(x + 2) < 0$ if and only if either $(x - 1 < 0$ and $x + 2 > 0)$ or $(x - 1 > 0$ and $x + 2 < 0)$ if and only if either $(-2 < x < 1)$ or $(x > 1$ and $x < -2)$. The last possibility can't occur, so $(x - 1)(x + 2) < 0$ if and only if $-2 < x < 1$.

(c) $x^2 + 4x - 21 = (x + 7)(x - 3)$, so $x^2 + 4x - 21 > 0$ if and only if either $(x + 7 > 0$ and $x - 3 > 0)$ or $(x + 7 < 0$ and $x - 3 < 0)$ if and only if either $x > 3$ or $x < -7$.

(d) $2x^2 + x < 3$ if and only if $2x^2 + x - 3 < 0$ if and only if $(x - 1)(2x + 3) < 0$ if and only if either $(x - 1 < 0$ and $2x + 3 > 0)$ or $(x - 1 > 0$ and $2x + 3 < 0)$ if and only if either $(-3/2 < x < 1)$ or $(x > 1$ and $x < -3/2)$. The last possibility can't occur, so $-3/2 < x < 1$.

(e) $4x^2 + 10x - 6 < 0$ if and only if $2(x + 3)(2x - 1) < 0$ if and only if either $(x + 3 < 0$ and $2x - 1 > 0)$ or $(x + 3 > 0$ and $2x - 1 < 0)$ if and only if either $(x < -3$ and $x > 1/2)$ or $(-3 < x < 1/2)$. The penultimate possibility can't occur, so $-3 < x < 1/2$.

(f) $x^2 + 2x + 4 = (x + 1)^2 + 3 \geq 0 + 3 = 3 > 0$ for all real x.

5. (a) $-2 \leq x \leq 2$.

(b) $x \leq -3$ or $x \geq 3$.

(c) $x < 4/3$.

(d) $x < -3$ or $x > 4$.

6. (a) Because $x^2 + 4 > 0$ for all x, $x/(x^2 + 4)$ is positive if and only if $x > 0$.

(b) $x/(x^2 - 4) > 0$ if and only if $(x > 0$ and $x^2 - 4 > 0)$ or $(x < 0$ and $x^2 - 4 < 0)$ if and only if $x > 2$ or $-2 < x < 0$.

(c) $(x + 1)/(x - 3) > 0$ if and only if $(x + 1 > 0$ and $x - 3 > 0)$ or $(x + 1 < 0$ and $x - 3 < 0)$ if and only if $x > 2$ or $x < -1$.

(d) $\dfrac{x^2-1}{x^2-3x} = \dfrac{(x+1)(x-1)}{x(x-3)}$. The signs of the four factors depend on which of these intervals x is in: $(-\infty, -1)$, $(-1, 0)$, $(0, 1)$, $(1, 3)$, $(3, \infty)$. For example, if x is in $(-1, 0)$, then $x+1$ is positive while $x-1$, x, and $x-3$ are negative, so $\dfrac{(x+1)(x-1)}{x(x-3)} < 0$. Checking the other intervals in this way, we find that $\dfrac{x^2-1}{x^2-3x}$ is positive if and only if either $x < -1$ or $0 < x < 1$ or $x > 3$.

7. (a) All values of a.
   (b) No values of a.

8. $a = b$.

9. (a) Yes.
   (b) No. If $a = b$, then $a < b$ is false.

10. Two points lie on a horizontal line if and only if they have the same y-coordinate, and on a vertical line if and only if they have the same x-coordinate.
    (a) Vertical.
    (b) Horizontal.
    (c) Horizontal.
    (d) Vertical.
    (e) Horizontal.
    (f) Vertical.
    (g) Vertical.
    (h) Horizontal.

11. $(3, 2)$

12. (a) $\sqrt{(1-6)^2+(2-7)^2} = \sqrt{25+25} = \sqrt{50} = 5\sqrt{2}$.
    (b) $\sqrt{(2-(-1))^2+(5-3)^2} = \sqrt{9+4} = \sqrt{13}$.
    (c) $\sqrt{(-7-1)^2+(3-(-2))^2} = \sqrt{64+25} = \sqrt{89}$.
    (d) $\sqrt{(a-b)^2+(b-a)^2} = \sqrt{2(a-b)^2} = \sqrt{2}\,|a-b|$.

13. (a) $x < 2$

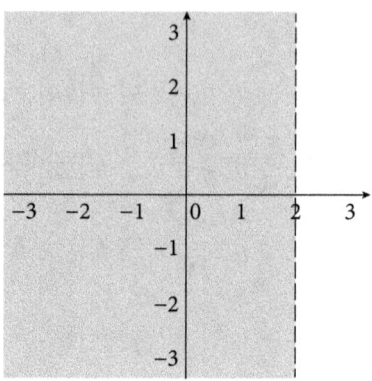

(b) $-1 < y \le 2$

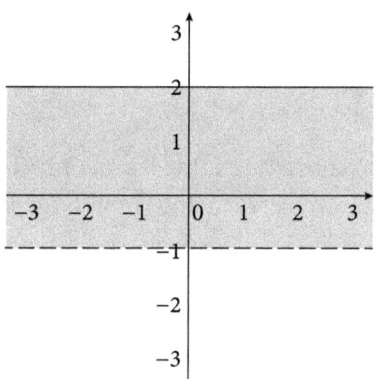

(c) $0 \le x \le 1$ and $0 \le y \le 1$

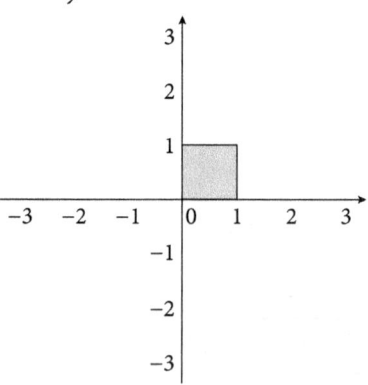

(d) $x = -1$

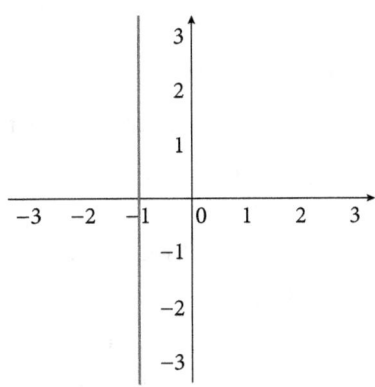

(e) $y = 3$

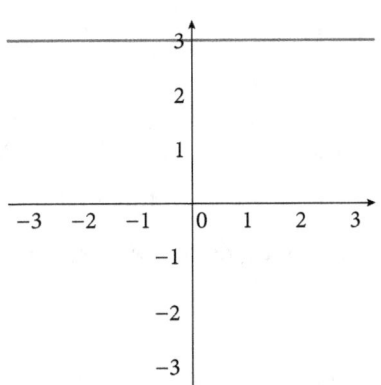

(f) $x = y$

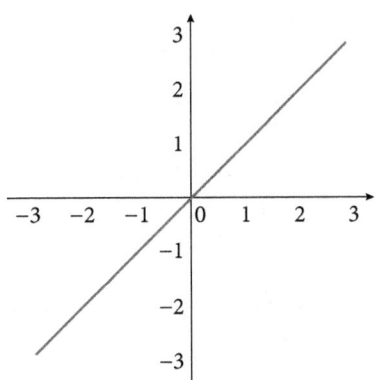

14. The distances from (6, 5) to (−2, 1) and to (2, −3) are, respectively, $\sqrt{(6+2)^2 + (5-1)^2} = \sqrt{80}$ and $\sqrt{(6-2)^2 + (5+3)^2} = \sqrt{80}$. Because these values are equal, (6, 5) is on the perpendicular bisector.

15. The center is the midpoint of (2, −2) and (−6, 5); that is, (−2, 3/2). The radius is half the diameter; that is $\frac{1}{2}\sqrt{(2+6)^2 + (-2-5)^2} = \frac{1}{2}\sqrt{113}$.

16. If the point is (x, y), then the three distances are $\sqrt{(x+9)^2 + y^2}$, $\sqrt{(x-6)^2 + (y-3)^2}$, and $\sqrt{(x+5)^2 + (y-6)^2}$. Hence $(x + 9)^2 + y^2 = (x − 6)^2 + (y − 3)^2 = (x + 5)^2 + (y − 6)^2$. Subtracting $x^2 + y^2$ throughout yields $18x + 81 = −12x − 6y + 45 = 10x − 12y + 61$. From $18x + 81 = −12x − 6y + 45$ we obtain $y = −5x − 6$. Substituting this into $18x + 81 = 10x − 12y + 61$ then gives $x = −1$. Hence $y = −5(−1) − 6 = −1$ and the point is (−1, −1).

17. The points (a, b) and (b, a) are symmetric with respect to the line $x = y$. (For any value of x, the distances from (x, x) to the given points are equal. Hence $x = y$ is the perpendicular bisector.)

18. Let x be the length of one side of the isosceles right triangle. Then $h^2 = x^2 + x^2 = 2x^2$, so $x = \dfrac{h}{\sqrt{2}}$.

19. Consider the triangles DBC and ABD in the following figure.

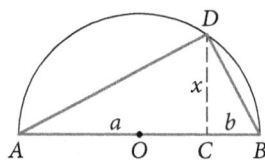

Both angles CDB and CAD are complements of angle ABD; hence $\angle CBD = \angle CAD$. Also, $\angle DCB = \angle ACD$ because both are right angles. Thus triangles DCB and ACD are similar. Therefore $\dfrac{a}{x} = \dfrac{x}{b}$, so $x = \sqrt{ab}$. But x is clearly less than or equal to the radius of the circle; that is, $\sqrt{ab} \leq \dfrac{a+b}{2}$.

20. Note that rule 3 in "Inequalities" on page 3 is also true if "<" is replaced by "≤". For, if $a \leq b$, then either $a = b$ or $a < b$. Hence either $a + c = b + c$ or $a + c < b + c$; that is, $a + c \leq b + c$.

From $a \leq b$ we get $a + c \leq b + c$, and from $c \leq d$ we get $b + c \leq b + d$. Hence $a + c \leq b + d$.

Because $a \leq |a|$ and $b \leq |b|$, $a + b \leq |a| + |b|$. Because $-|a| \leq a$ and $-|b| \leq b$, $-|a| - |b| \leq a + b$. Adding $|a| + |b| - a - b$ to both sides of this inequality gives $-a - b \leq |a| + |b|$; that is, $-(a + b) \leq |a| + |b|$. But $|a + b|$ is either $a + b$ or $-(a + b)$; hence it is less than or equal to $|a| + |b|$.

21. We have

$$x_{n+1} = \frac{1}{2}\left(x_n + \frac{2}{x_n}\right).$$

Starting with $x_1 = 1$, we obtain (to five decimal places) $x_2 = 1.50000$, $x_3 \approx 1.41667$, $x_4 \approx 1.41422$, $x_5 \approx 1.41421$, $x_6 \approx 1.41421$. If we start with $x_1 = 3/2 = 1.5$, then we begin one step further into the iterations that begin with 1. Hence $x_2 \approx 1.41667$, and so on.

22. Let n be an integer such that $n > 1/(b - a)$. Let m be the largest integer such that $m < nb$. Note that $m/n < b$. We claim that $m/n > a$, so m/n is the desired rational number. First observe that, because $n > 1/(b - a)$, we have $n(b - a) > 1$, so $na < nb - 1$. So, if $m/n \leq a$, then $m \leq na < nb - 1$ and $m + 1 < nb$, violating the choice of m. Hence $m/n > a$. Because

$$a < \frac{a+b}{2} < \frac{a+2b}{3} < \frac{a+3b}{4} < \cdots < b,$$

there exist rational numbers $c_0, c_1, \ldots$ with

$$a < c_0 < \frac{a+b}{2} < c_1 < \frac{a+2b}{3} < c_2 < \frac{a+3b}{4} < c_3 < \cdots < b.$$

Hence $c_0, c_1, \ldots$ are infinitely many rationals between a and b.

23. No, neither the sum nor the product of irrational numbers need be irrational. Let $a = \sqrt{2}$ and $b = -\sqrt{2}$. Then a and b are irrational, but $a + b = 0$ is rational and $ab = -2$ is rational.

24. Note that triangles ADB and DCB are similar, because $\angle ADB = \angle DCB = \pi/2$ and $\angle ABD = \angle DBC$.

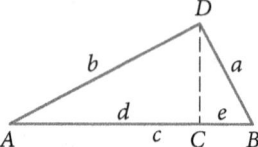

Hence

$$\frac{a}{c} = \frac{DB}{AB} = \frac{CB}{DB} = \frac{e}{a}.$$

Similarly, triangles ADB and ACD are similar, so

$$\frac{b}{c} = \frac{AD}{AB} = \frac{AC}{AD} = \frac{d}{b}.$$

Clearing denominators, $a^2 = ce$ and $b^2 = cd$, so $a^2 + b^2 = ce + cd = c(e + d) = c^2$.

25. It's clear from the following figure that x divides the interval $[x_1, x_2]$ (or $[x_2, x_1]$ if $x_2 < x_1$) in the ratio $q{:}p$; that is, $\dfrac{x - x_1}{x_2 - x} = \dfrac{q}{p}$.

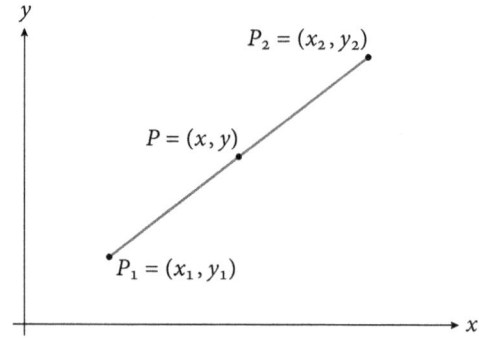

Hence $p(x - x_1) = q(x_2 - x)$, $px - px_1 = qx_2 - qx$, $(p + q)x = px_1 + qx_2$, and $x = \dfrac{px_1 + qx_2}{p + q}$. Similarly, $y = \dfrac{py_1 + qy_2}{p + q}$.

Chapter 2

1. (a)

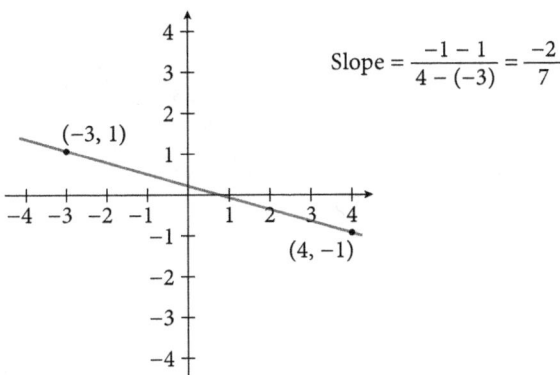

$$\text{Slope} = \frac{-1 - 1}{4 - (-3)} = \frac{-2}{7}$$

(b)

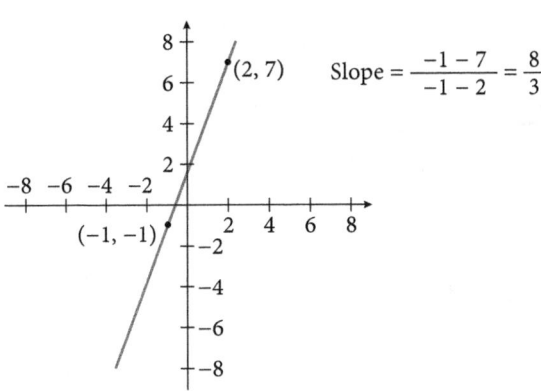

$$\text{Slope} = \frac{-1 - 7}{-1 - 2} = \frac{8}{3}$$

(c)

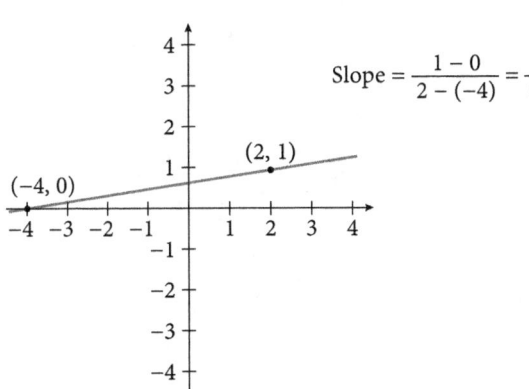

$$\text{Slope} = \frac{1 - 0}{2 - (-4)} = \frac{1}{6}$$

(d)

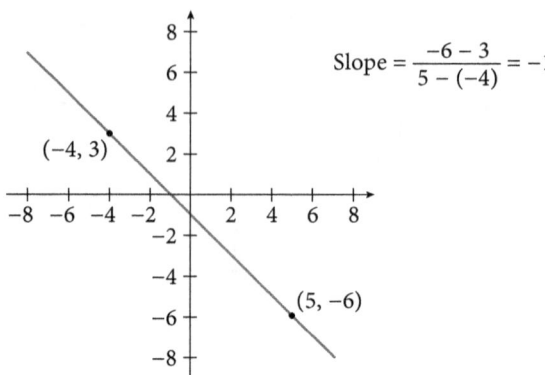

Slope $= \dfrac{-6-3}{5-(-4)} = -1$

$(-4, 3)$

$(5, -6)$

(e)

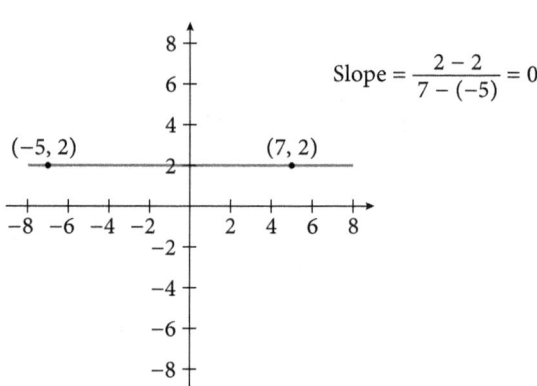

Slope $= \dfrac{2-2}{7-(-5)} = 0$

$(-5, 2)$

$(7, 2)$

(f)

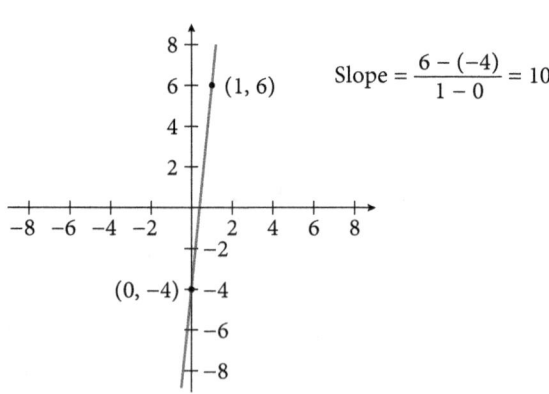

Slope $= \dfrac{6-(-4)}{1-0} = 10$

$(1, 6)$

$(0, -4)$

2. If segments AB and BC have the same slope, then points A, B, and C are collinear (lie in the same straight line). They can't be located in different lines because they share the same point B. Note that B doesn't have to be the "middle" point—you can choose any two pairs of points for slope calculation, and they will have a point in common. For the points $A = (1, 1)$, $B = (-5, -2)$, and $C = (5, 3)$, the slopes of AB and BC are

$$m_{AB} = \frac{-2-1}{-5-1} = \frac{1}{2} \quad \text{and} \quad m_{BC} = \frac{3-(-2)}{5-(-5)} = \frac{1}{2}.$$

The slopes m_{AB} and m_{BC} are equal, so A, B, and C are collinear.

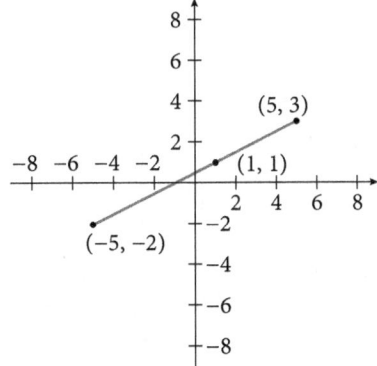

3. We will compute the slopes of the four sides to show that a and c are parallel, b and d are parallel, and a is perpendicular to b.

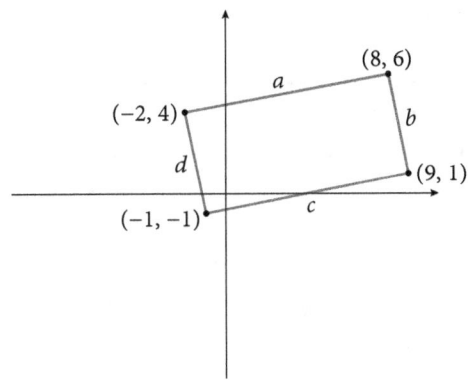

The slopes of a, b, c, and d are

$$m_a = \frac{6-4}{8-(-2)} = \frac{1}{5}, \qquad m_b = \frac{1-6}{9-8} = -5,$$

$$m_c = \frac{1-(-1)}{9-(-1)} = \frac{1}{5}, \qquad m_d = \frac{-1-4}{-1-(-2)} = -5.$$

Hence a is parallel to c and b is parallel to d. Finally, because $(1/5) \times (-5) = -1$, a is perpendicular to b.

Note that to show whether four points are the vertices of a *square*, you must perform the steps above *and* use the distance formula (page 8) to show that all four sides are the same length.

4. (a)

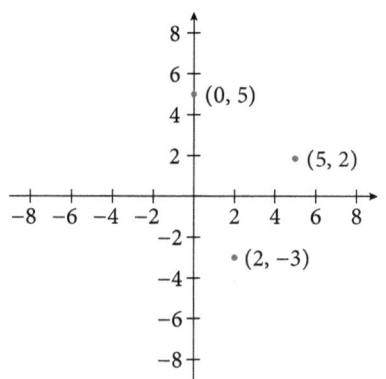

The slopes of the sides are

$$\frac{2+3}{5-2} = \frac{5}{3}, \qquad \frac{5+3}{0-2} = -4, \qquad \frac{5-2}{0-5} = -\frac{3}{5}.$$

Because 5/3 and −3/5 are negative reciprocals, the points form a right triangle.

(b)

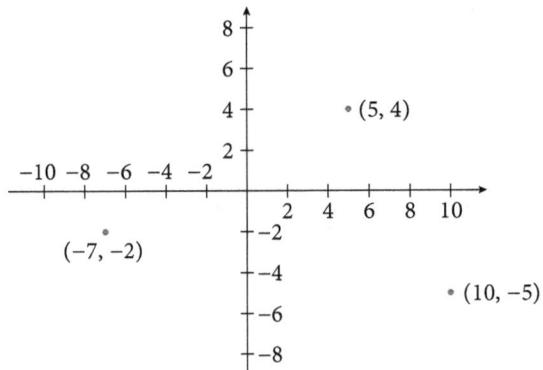

The slopes of the sides are

$$\frac{4-(-5)}{5-10}=-\frac{9}{5}, \quad \frac{-2-(-5)}{-7-10}=-\frac{3}{17}, \quad \frac{-2-4}{-7-5}=\frac{1}{2}.$$

No two of these slopes are negative reciprocals, so the points don't form a right triangle.

(c)

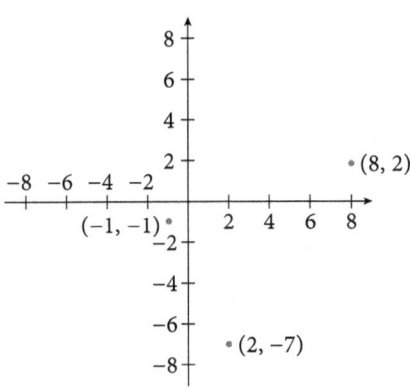

The slopes of the sides are

$$\frac{-1-2}{-1-8}=\frac{1}{3}, \quad \frac{-7-2}{2-8}=\frac{3}{2}, \quad \frac{-7-(-1)}{2-(-1)}=-2.$$

No two of these slopes are negative reciprocals, so the points don't form a right triangle.

(d)

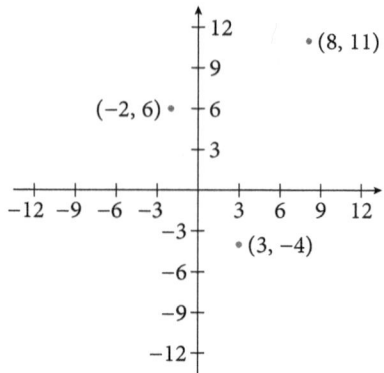

The slopes of the sides are

$$\frac{-4-6}{3-(-2)} = -2, \quad \frac{11-6}{8-(-2)} = \frac{1}{2}, \quad \frac{11-(-4)}{8-3} = 3.$$

Because -2 and $\frac{1}{2}$ are negative reciprocals, the points form a right triangle.

5. (a) y-intercept $= 1/7$; $y = (-2/7)x + 1/7$.
    (b) y-intercept $= 5/3$; $y = (8/3)x + 5/3$.
    (c) y-intercept $= 2/3$; $y = (1/6)x + 2/3$.
    (d) y-intercept $= -1$; $y = -x - 1$.
    (e) y-intercept $= 2$; $y = 2$.
    (f) y-intercept $= -4$; $y = 10x - 4$.

6. (a) $y + 3 = -4(x - 2)$; that is, $y = -4x + 5$.
    (b) The slope is $\dfrac{-1-2}{3-(-4)} = -\dfrac{3}{7}$, so the equation is $y - 2 = -\dfrac{3}{7}(x + 4)$
    or $3x + 7y - 2 = 0$.
    (c) Plug in the slope and intercept: $y = \dfrac{2}{3}x - 4$.
    (d) The line is horizontal: $y = -4$.
    (e) The line is vertical: $x = 1$.

(f) $x + 3y = 7$ can be rewritten as $y = -\dfrac{1}{3}x + \dfrac{7}{3}$, so it has slope –1/3. The desired line also must have slope –1/3, so its equation is

$y + 2 = -\dfrac{1}{3}(x - 4)$, or $x + 3y + 2 = 0$.

(g) $y + 7 = 2x$ has slope 2, so the desired line has slope –½. Its equation is therefore $y - 3 = -\dfrac{1}{2}(x - 5)$, or $x + 2y = 11$.

(h) The line through (–2, –2) and (1, 0) has slope $\dfrac{0 - (-2)}{1 - (-2)} = \dfrac{2}{3}$. Hence the desired line is $y - 3 = \dfrac{2}{3}(x + 4)$, or $2x - 3y = -17$.

(i) The slope of the segment joining (1, –1) and (5, 7) is $\dfrac{7 - (-1)}{5 - 1} = 2$, so the desired line has slope –½. It must pass through the midpoint of (1, –1) and (5, 7), namely, $\left(\dfrac{1}{2}(1 + 5), \dfrac{1}{2}(-1 + 7) \right) = (3, 3)$. Thus its equation is $y - 3 = -\dfrac{1}{2}(x - 3)$, or $x + 2y = 9$.

(j) The slope is $\tan 135° = -1$. Hence the equation is $y - 3 = -(x + 2)$, or $y = 1 - x$, or $x + y = 1$.

7. (a) $5x + 3y + 15 = 0$, so $5x + 3y = -15$, and $\dfrac{x}{-3} + \dfrac{y}{-5} = 1$.

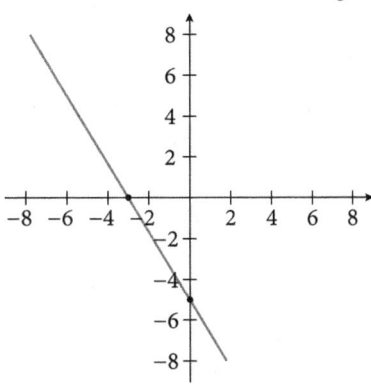

(b) $3x = 8y - 24$, so $3x - 8y = -24$, and $\dfrac{x}{-8} + \dfrac{y}{3} = 1.$

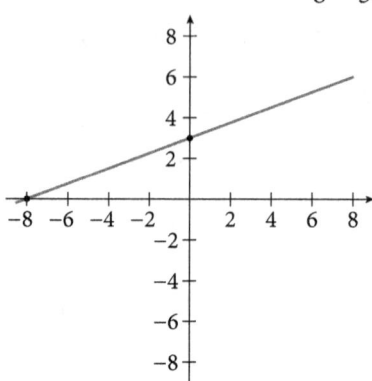

(c) $y = 6 - 6x$, so $6x + y = 6$ and $\dfrac{x}{1} + \dfrac{y}{6} = 1.$

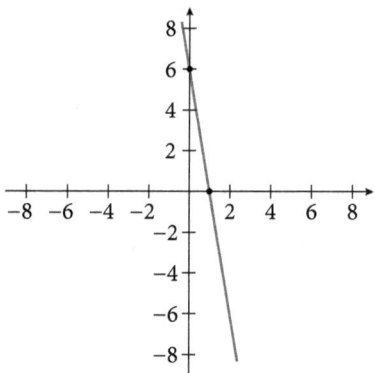

(d) $2x - 3y = 9$, so $\dfrac{x}{9/2} + \dfrac{y}{-3} = 1.$

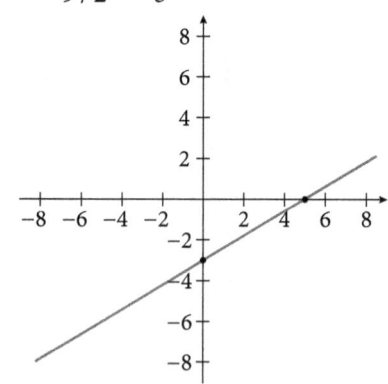

8. Adding the first equation, $3x + 4y = 7$, to twice the second equation, $2x - 4y = 12$, gives $5x = 19$, so $x = 19/5$. Subtracting three times the second equation from the first gives $10y = -11$, so $y = -11/10$. The intersection is $(19/5, -11/10)$.

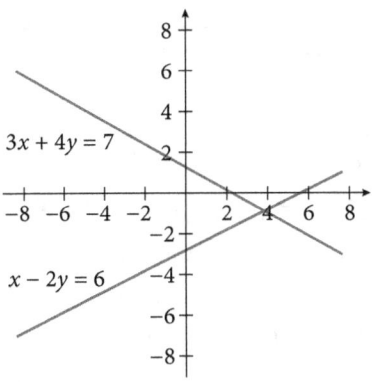

9. Taking the C-axis as horizontal and the F-axis as vertical, the slope is $(212 - 32)/(100 - 0) = 9/5$. The F-intercept is 32, so the equation is $F = (9/5)C + 32$.

10. Let the vertices of the quadrilateral, in cyclic order, be (x_0, y_0), (x_1, y_1), (x_2, y_2), (x_3, y_3). Then the midpoints are

$$M_0 = \left(\frac{x_0 + x_1}{2}, \frac{y_0 + y_1}{2} \right), \quad M_1 = \left(\frac{x_1 + x_2}{2}, \frac{y_1 + y_2}{2} \right),$$

$$M_2 = \left(\frac{x_2 + x_3}{2}, \frac{y_2 + y_3}{2} \right), \quad M_3 = \left(\frac{x_3 + x_0}{2}, \frac{y_3 + y_0}{2} \right).$$

The slope of $M_0 M_1$ is

$$\frac{\frac{1}{2}(y_1 + y_2) - \frac{1}{2}(y_0 + y_1)}{\frac{1}{2}(x_1 + x_2) - \frac{1}{2}(x_0 + x_1)} = \frac{y_2 - y_0}{x_2 - x_0}.$$

The slope of $M_3 M_2$ is

$$\frac{\frac{1}{2}(y_2 + y_3) - \frac{1}{2}(y_3 + y_0)}{\frac{1}{2}(x_2 + x_3) - \frac{1}{2}(x_3 + x_0)} = \frac{y_2 - y_0}{x_2 - x_0}.$$

Because these two slopes are equal, M_0M_1 is parallel to M_3M_2. Similarly, M_1M_2 is parallel to M_0M_3, so $M_0M_1M_2M_3$ is a parallelogram.

11. The length of the horizontal side is a. The lengths of the other two sides are

$$\sqrt{(b-0)^2 + (c-0)^2} = \sqrt{b^2 + c^2}$$

and

$$\sqrt{(b-a)^2 + (c-0)^2} = \sqrt{(b-a)^2 + c^2}.$$

Hence we have $a^2 = (b^2 + c^2) + [(b-a)^2 + c^2]$, which simplifies to $ab = b^2 + c^2$. The slopes of the nonhorizontal sides are

$$\frac{c-0}{b-0} = \frac{c}{b} \quad \text{and} \quad \frac{c-0}{b-a} = \frac{c}{b-a}.$$

Their product is

$$\frac{c^2}{b(b-a)} = \frac{c^2}{b^2 - ab} = \frac{c^2}{b^2 - (b^2 + c^2)} = \frac{c^2}{-c^2} = -1.$$

Hence the sides are perpendicular.

12. Let the given point be (x_0, y_0). Because this point is on the line, we have

$$y_0 - y_1 = \frac{y_2 - y_1}{x_2 - x_1}(x_0 - x_1).$$

The point-slope equation is

$$y - y_0 = \frac{y_2 - y_1}{x_2 - x_1}(x - x_0).$$

Adding the equal quantities $y_0 - y_1$, and

$$\frac{y_2 - y_1}{x_2 - x_1}(x_0 - x_1)$$

to the two sides of this equation, we see that it's equivalent to

$$y - y_1 = \frac{y_2 - y_1}{x_2 - x_1}(x - x_1).$$

Because this equation is independent of (x_0, y_0), the point-slope equation is the same no matter which point is used.

13. A line passing through the vertex of a triangle which is perpendicular to the opposite side is called an **altitude**.

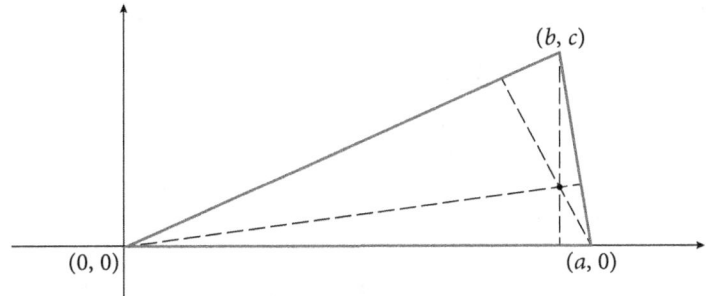

The altitude through (b, c) is vertical; its equation is $x = b$. The slope of the side opposite $(0, 0)$ is $c/(b - a)$, so the slope of the altitude through $(0, 0)$ is $(a - b)/c$. Its equation is

$$y = \frac{a-b}{c} x.$$

The slope of the side opposite $(a, 0)$ is c/b, so the slope of the altitude through $(a, 0)$ is $-b/c$. Its equation is

$$\frac{y}{x-a} = -\frac{b}{c};$$

that is,

$$y = -\frac{b}{c}(x-a).$$

The intersection of the three lines is the point

$$\left(b, \frac{b}{c}(a-b) \right).$$

14. Let P be the point on the line closest to (x_0, y_0). Then the line through P and (x_0, y_0) is perpendicular to the given line. Because the given line has slope $-A/B$, the perpendicular has slope B/A; its equation is

$$y - y_0 = \frac{B}{A}(x - x_0).$$

So P satisfies both $Ax + By + C = 0$ and

$$y - y_0 = \frac{B}{A}(x - x_0).$$

Hence

$$P = \left(\frac{B^2 x_0 - ABy_0 - AC}{A^2 + B^2}, \frac{A^2 y_0 - ABx_0 - BC}{A^2 + B^2} \right).$$

The square of the distance from (x_0, y_0) to P is

$$\left(x_0 - \frac{B^2 x_0 - ABy_0 - AC}{A^2 + B^2} \right)^2 + \left(y_0 - \frac{A^2 y_0 - ABx_0 - BC}{A^2 + B^2} \right)^2 =$$

$$\frac{1}{(A^2 + B^2)^2} \left[(A^2 x_0 + ABy_0 + AC)^2 + (B^2 y_0 + ABx_0 + BC)^2 \right] =$$

$$\frac{1}{(A^2 + B^2)^2} \left[A^2 (Ax_0 + By_0 + C)^2 + B^2 (Ax_0 + By_0 + C)^2 \right] =$$

$$\frac{1}{A^2 + B^2} (Ax_0 + By_0 + C)^2.$$

Hence the distance is

$$\frac{|Ax_0 + By_0 + C|}{\sqrt{A^2 + B^2}}.$$

15. (a) By Problem 14 in Chapter 2, the distance from (x, y) to $3x + 4y - 10 = 0$ is

$$\frac{|3x + 4y - 10|}{\sqrt{3^2 + 4^2}} = \frac{1}{5} |3x + 4y - 10|.$$

The distance from (x, y) to $4x - 3y - 5 = 0$ is

$$\frac{|4x - 3y - 5|}{\sqrt{4^2 + (-3)^2}} = \frac{1}{5} |4x - 3y - 5|.$$

Hence

$$\frac{1}{5} |3x + 4y - 10| = \frac{1}{5} |4x - 3y - 5|,$$

so either $3x + 4y - 10 = 4x - 3y - 5$ or $3x + 4y - 10 = -(4x - 3y - 5)$. These cases reduce to

$$y = \frac{1}{7}(x+5)$$

and $y = 15 - 7x$, respectively.

(b) The distance from (x, y) to $y = 0$ is $|y|$. Because $x = y$ if and only if $x - y + 0 = 0$, Problem 14 in Chapter 2 implies that the distance from (x, y) to $x = y$ is

$$\frac{|x-y+0|}{\sqrt{1^2 + (-1)^2}} = \frac{1}{\sqrt{2}}|x-y|,$$

so either

$$y = \frac{1}{\sqrt{2}}(x-y) \quad \text{or} \quad y = -\frac{1}{\sqrt{2}}(x-y).$$

These equations reduce to $x = (1 + \sqrt{2})y$ and $x = (1 - \sqrt{2})y$.

Chapter 3

1. Simply plug into the formula $(x - h)^2 + (y - k)^2 = r^2$.
   (a) $(x - 4)^2 + (y - 6)^2 = 9$.
   (b) $(x + 3)^2 + (y - 7)^2 = 5$.
   (c) $(x + 5)^2 + (y + 9)^2 = 49$.
   (d) $(x - 1)^2 + (y + 6)^2 = 2$.
   (e) $(x - a)^2 + (y - 0)^2 = a^2$; thus $x^2 - 2ax + a^2 + y^2 = a^2$, so
   $x^2 + y^2 = 2ax$.
   (f) $(x - 0)^2 + (y - a)^2 = a^2$; thus $x^2 + y^2 = 2ay$.

2. (a) We know the circle's center $(h, k) = (-7, 3)$. To find its radius r, use the distance formula (page 8):

$$r = \sqrt{(-7-4)^2 + (3-(-1))^2} = \sqrt{137}.$$

Simply plug into the formula $(x - h)^2 + (y - k)^2 = r^2$. The circle's equation is

$$(x + 7)^2 + (y - 3)^2 = 137.$$

(b) We need to find the circle's center (h, k) and radius r. To find the center, which lies equidistant from the two given points, use the midpoint formulas (page 10):

$$(h,k) = \left(\frac{x_1 + x_2}{2}, \frac{y_1 + y_2}{2} \right) = \left(\frac{-1+7}{2}, \frac{1+9}{2} \right) = (3,5).$$

Now we can use the distance formula to find the radius:

$$r = \sqrt{(3-7)^2 + (5-9)^2} = \sqrt{32}.$$

Simply plug into the formula $(x - h)^2 + (y - k)^2 = r^2$. The circle's equation is

$$(x - 3)^2 + (y - 5)^2 = 32.$$

(c) We know the circle's center $(h, k) = (-3, -5)$ and need to find its radius r. By Problem 14 in Chapter 2, the distance from the center $(-3, -5) = (x_0, y_0)$ to the tangent line $12x + 5y - 4 = 0$ is the radius:

$$r = \frac{|Ax_0 + By_0 + C|}{\sqrt{A^2 + B^2}} = \frac{|12(-3) + 5(-5) + (-4)|}{\sqrt{12^2 + 5^2}} = \frac{65}{13} = 5.$$

Simply plug into the formula $(x - h)^2 + (y - k)^2 = r^2$. The circle's equation is

$$(x + 3)^2 + (y + 5)^2 = 25.$$

3. (a) $x^2 + y^2 - 4x - 4y = 0$,
$x^2 - 4x + 4 + y^2 - 4y + 4 = 8$,
$(x - 2)^2 + (y - 2)^2 = 8$.
The graph is a circle of radius $\sqrt{8}$ centered at $(2, 2)$.

(b) $x^2 + y^2 - 18x - 14y + 130 = 0$,
$x^2 - 18x + 81 + y^2 - 14y + 49 = 0$,
$(x - 9)^2 + (y - 7)^2 = 0$.
The graph is the point $(9, 7)$ (a circle of radius zero).

(c) $x^2 + y^2 + 8x + 10y + 40 = 0$,
$x^2 + 8x + 16 + y^2 + 10y + 25 = 1$,
$(x + 4)^2 + (y + 5)^2 = 1$.
The graph is a circle of radius 1 centered at $(-4, -5)$.

(d) $4x^2 + 4y^2 + 12x - 32y + 37 = 0$,
$x^2 + 3x + 9/4 + y^2 - 8y + 16 + 37/4 = 9/4 + 16$,
$(x + 3/2)^2 + (y - 4)^2 = 9$.
The graph is a circle of radius 3 centered at $(-3/2, 4)$.

(e) $x^2 + y^2 - 8x + 12y + 53 = 0$,
$x^2 - 8x + 16 + y^2 + 12y + 36 = -1$,
$(x - 4)^2 + (y + 6)^2 = -1$.
An empty graph; a radius of $\sqrt{-1}$ is impossible.

(f) $x^2 + y^2 - \sqrt{2}x + \sqrt{2}y + 1 = 0$,
$x^2 - \sqrt{2}x + \frac{1}{2} + y^2 + \sqrt{2}y + \frac{1}{2} = 0$,
$(x - 1/\sqrt{2})^2 + (y + 1/\sqrt{2})^2 = 0$.
The graph is the point $(1/\sqrt{2}, -1/\sqrt{2})$ (a circle of radius 0).

(g) $x^2 + y^2 - 16x + 6y - 48 = 0$,
$x^2 - 16x + 64 + y^2 + 6y + 9 = 121$,
$(x - 8)^2 + (y + 3)^2 = 121.$
The graph is a circle of radius 11 centered at $(8, -3)$.

4.
$$ax^2 + bx + c = 0$$
$$x^2 + \frac{b}{a}x + \frac{c}{a} = 0$$
$$x^2 + \frac{b}{a}x = -\frac{c}{a}$$
$$x^2 + \frac{b}{a}x + \frac{b^2}{4a^2} = -\frac{c}{a} + \frac{b^2}{4a^2}$$
$$\left(x + \frac{b}{2a}\right)^2 = \frac{b^2 - 4ac}{4a^2}$$
$$x + \frac{b}{2a} = \frac{\pm\sqrt{b^2 - 4ac}}{2a}$$
$$x = \frac{-b \pm \sqrt{b^2 - 4ac}}{2a}$$

The equation has distinct real roots if $b^2 - 4ac > 0$, equal real roots if $b^2 - 4ac = 0$, and no real roots if $b^2 - 4ac < 0$.

5. Substituting $y = mx + 4$ in $x^2 + y^2 = 2y$ gives $(1 + m^2)x^2 + 6mx + 8 = 0$, which has exactly one real root if and only if $(6m)^2 - 4(1 + m^2)8 = 0$; that is, if and only if $m = \sqrt{8}$ or $m = -\sqrt{8}$. The equations of the tangent lines are $y = \sqrt{8}x + 4$ and $y = -\sqrt{8}x + 4$. (Note that the vertical line through $(0, 4)$, which doesn't have the form $y = mx + 4$, is not tangent to the circle; it intersects the circle at $(0, 0)$ and $(0, 2)$.)

6. (a) $y^2 = 12x$; focus $(3, 0)$; vertex $(0, 0)$; directrix $x = -3$.

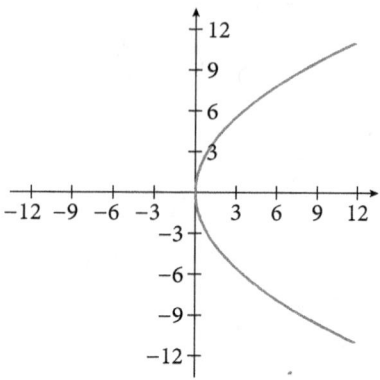

(b) $y = 4x^2$; focus $(0, 1/16)$; vertex $(0, 0)$; directrix $y = -1/16$.

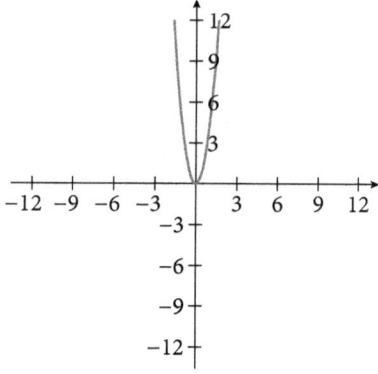

(c) $2x^2 + 5y = 0$; focus $(0, -5/8)$; vertex $(0, 0)$; directrix $y = 5/8$.

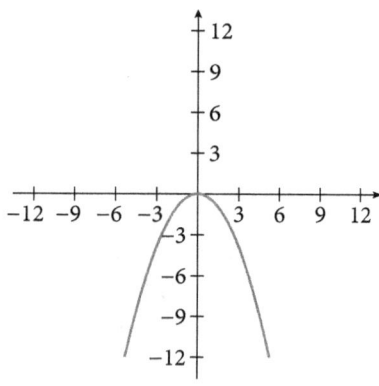

(d) $4x + 9y^2 = 0$; focus $(-1/9, 0)$; vertex $(0, 0)$; directrix $x = 1/9$.

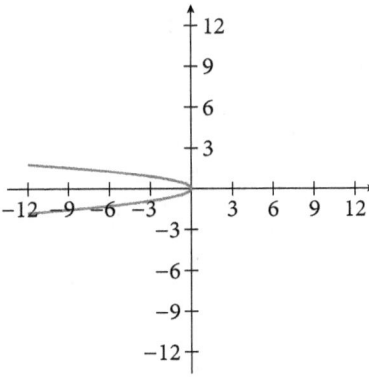

(e) $x = -2y^2$; focus $(-1/8, 0)$; vertex $(0, 0)$; directrix $x = 1/8$.

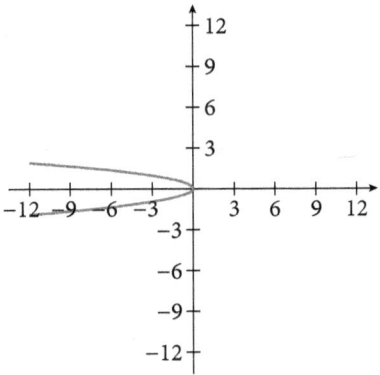

(f) $12y = -x^2$; focus $(0, -3)$; vertex $(0, 0)$; directrix $y = 3$.

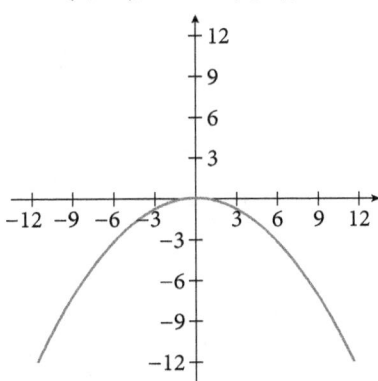

(g) $16y^2 = x$; focus $(1/64, 0)$; vertex $(0, 0)$; directrix $x = -1/64$.

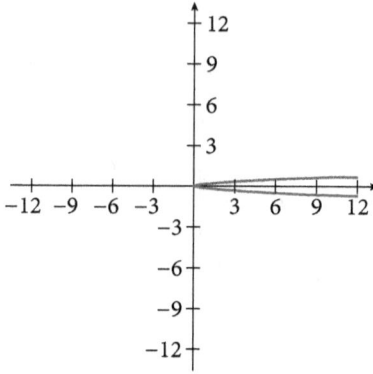

(h) $24x^2 = y$; focus $(0, 1/96)$; vertex $(0, 0)$; directrix $y = -1/96$.

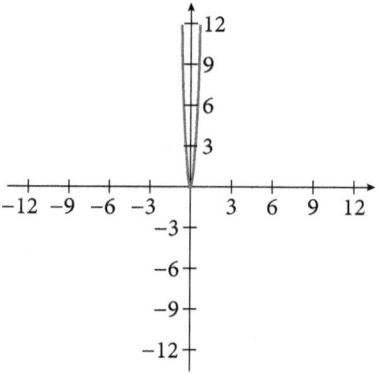

(i) $y^2 + 8y - 16x = 16$; focus $(2, -4)$; vertex $(-2, -4)$; directrix $x = -6$.

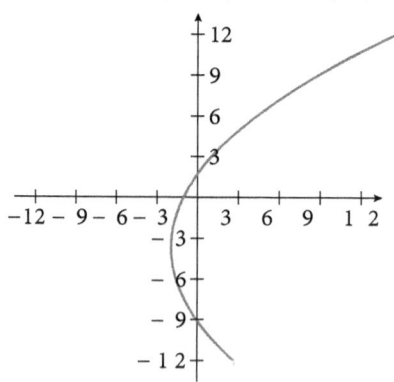

(j) $x^2 + 2x + 29 = 7y$; focus $(-1, 23/4)$; vertex $(-1, 4)$; directrix $y = 9/4$.

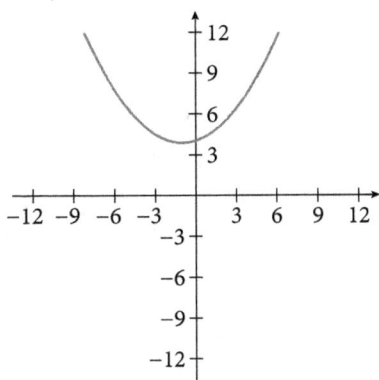

7. (a) $y^2 = -12x$.

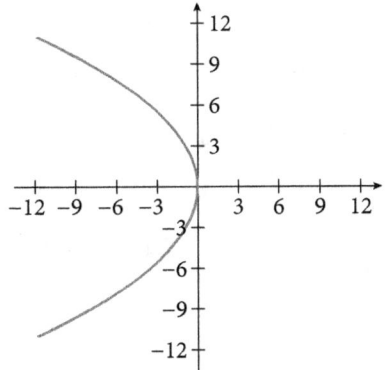

(b) $x^2 = 4y$.

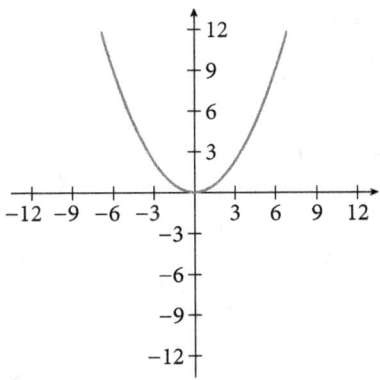

(c) $y^2 = 8x$.

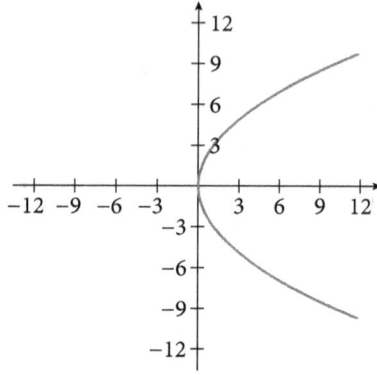

(d) $x^2 = (-4/3)y$.

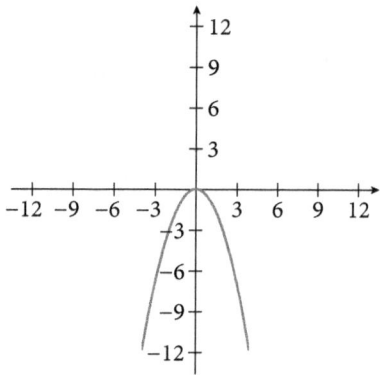

(e) The distances from (x, y) to the directrix and the focus are, respectively, $|x - 2|$ and $\sqrt{(x+4)^2 + y^2}$. So the equation is $|x - 2| = \sqrt{(x+4)^2 + y^2}$, which, upon squaring, becomes $(x - 2)^2 = (x + 4)^2 + y^2$. Thus we have $x^2 - 4x + 4 = x^2 + 8x + 16 + y^2$, $0 = 12x + 12 + y^2$, or $y^2 + 12(x + 1) = 0$.

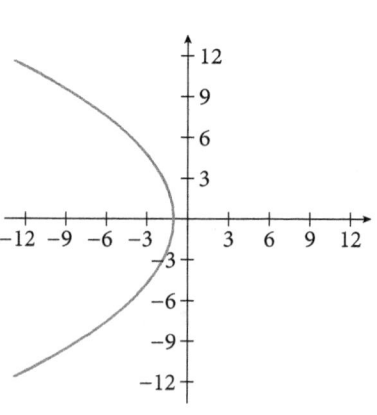

(f) The distances from (x, y) to the directrix and the focus are, respectively, $|y + 1|$ and $\sqrt{(x-3)^2 +(y-3)^2}$. So the equation is $|y+1| = \sqrt{(x-3)^2 +(y-3)^2}$. Hence the equation is $(y + 1)^2 = (x - 3)^2 + (y - 3)^2$, which simplifies to $8y = x^2 - 6x + 17$.

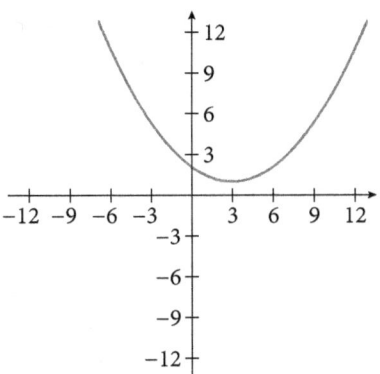

8. (a) $y = x^2 + 1$; focus $(0, 5/4)$; vertex $(0, 1)$; directrix $y = 3/4$.

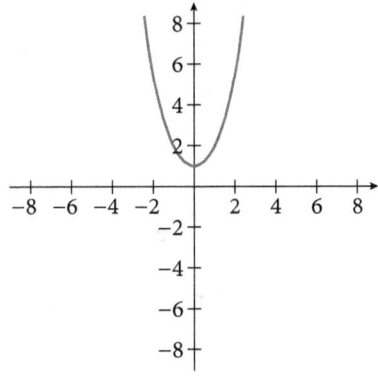

(b) $y = (x - 1)^2$; focus $(1, 1/4)$; vertex $(1, 0)$; directrix $y = -1/4$.

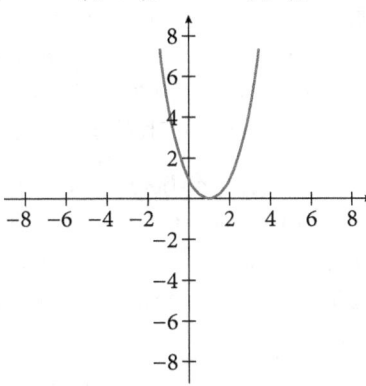

(c) $y = (x - 1)^2 + 1$; focus $(1, 5/4)$; vertex $(1, 1)$; directrix $y = 3/4$.

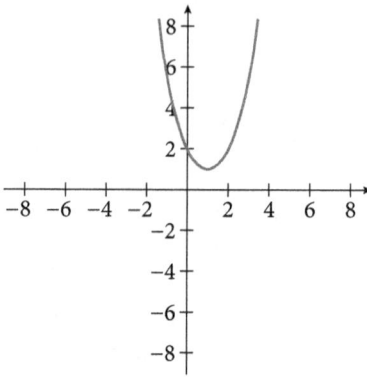

(d) $y = x^2 - x$; focus $(1/2, 0)$; vertex $(1/2, -1/4)$; directrix $y = -1/2$.

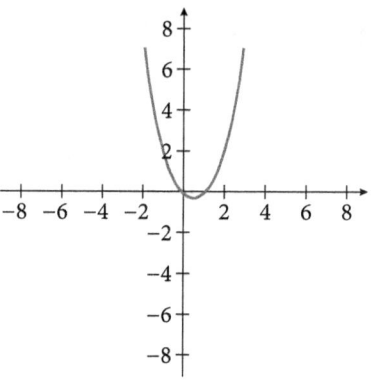

9. The curve has the form $y = kx^2$ for some k, where x is the horizontal distance from the nozzle in feet and y is the distance below the nozzle in feet. When $x = 10$, $y = 1$, so $k = 1/100$ and $y = x^2/100$. The water hits the ground when $y = 4$, so $x = 20$. The horizontal distance is 20 feet.

10. A nonvertical line is tangent to the parabola if and only if it meets the parabola in exactly one point. If the line's equation is $y = mx + b$, then the intersection point(s) must satisfy both $x^2 = 4py$ and $y = mx + b$, so $x^2 = 4p(mx + b)$, $x^2 - 4pmx - 4pb = 0$, and

$$x = \frac{4pm \pm \sqrt{16p^2m^2 + 16pb}}{2} = 2pm \pm 2\sqrt{p^2m^2 + pb}.$$

So there's exactly one intersection point if and only if $p^2m^2 + pb = 0$; that is, $b = -pm^2$. The tangent line's equation is $y = mx - pm^2$.

11. Choose coordinate axes and scale so that the focus of the parabola is at $(0, 1)$ and the directrix is $y = -1$. By equation (9), $x^2 = 4py$, the parabola's equation is $x^2 = 4y$. Note that every vertical line crosses the parabola. So if a line is tangent to the parabola, then it must have the form $y = mx + b$ and must meet the parabola only once. Substituting $y = mx + b$ into $x^2 = 4y$ yields $x^2 - 4mx - 4b = 0$, which has exactly one solution if and only if $(-4m)^2 - 4 \cdot 1(-4b) = 0$; that is, $b = -m^2$. Thus $y = mx + b$ is tangent to the parabola if and only if $b = -m^2$. Let $(a, -1)$ be a given point on the directrix. Then $y = mx - m^2$ passes through $(a, -1)$ if and only if $-1 = ma - m^2$; that is,

$$m = \frac{a \pm \sqrt{a^2 + 4}}{2}.$$

The product of the two slopes is

$$\frac{a + \sqrt{a^2 + 4}}{2} \cdot \frac{a - \sqrt{a^2 + 4}}{2} = -1,$$

so the tangents are perpendicular.

12. The line and the circle intersect if and only if the equation $x^2 + (3x + b)^2 = 4$ has a real root. This equation is equivalent to $10x^2 + 6bx + (b^2 - 4) = 0$, so its roots are real if and only if $(6b)^2 - 4 \cdot 10(b^2 - 4) \geq 0$; that is, if and only if $b^2 \leq 40$. Hence the line and the circle intersect for $-\sqrt{40} \leq b \leq \sqrt{40}$.

13. (a) $\sqrt{x^2 + y^2} = 2\sqrt{(x-a)^2 + y^2}$. Squaring and simplifying yields $3x^2 + 3y^2 - 8ax + 4a^2 = 0$. Completing the square, we find that $\left(x - \frac{4}{3}a\right)^2 + y^2 = \left(\frac{2}{3}a\right)^2$, which is the equation of a circle.

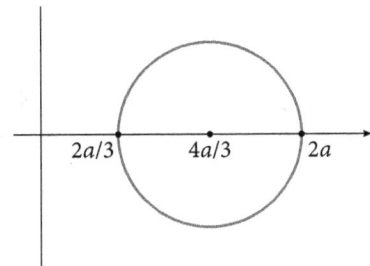

(b) $\sqrt{(x-a)^2 + y^2} \cdot \sqrt{(x+a)^2 + y^2} = a^2$. Squaring and simplifying, we get $(x^2 + y^2)^2 = 2a^2(x^2 - y^2)$.

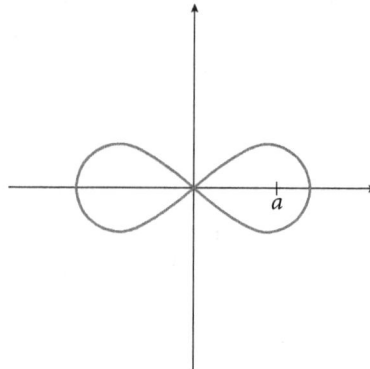

14. Choose a coordinate system so that the two points in question are $(0, 0)$ and $(a, 0)$. The equation of the locus is then

$$\sqrt{x^2 + y^2} = k\sqrt{(x-a)^2 + y^2}.$$

Squaring and rearranging produces $(k^2 - 1)x^2 - 2k^2ax + (k^2 - 1)y^2 + k^2a^2 = 0$. Dividing by $k^2 - 1$ and completing the square, we get

$$\left(x - \frac{k^2 a}{k^2 - 1}\right)^2 + y^2 = \left(\frac{ka}{k^2 - 1}\right)^2,$$

which is a circle of radius $|ka/(k^2 - 1)|$ centered at $(k^2a/(k^2 - 1), 0)$.

15. A line through $(1, 3)$ that's tangent to $x^2 + y^2 = 2$ can't be vertical (because $x = 1$ meets the circle at $(1, 1)$ and $(1, -1)$), so its equation has the form $y = ax + b$; that is, $ax - y + b = 0$. The line is tangent to the circle if and only if its distance from $(0, 0)$ is $\sqrt{2}$. By Problem 14 in Chapter 2, this means that

$$\frac{|b|}{\sqrt{a^2 + 1}} = \sqrt{2};$$

that is, $b^2 = 2(a^2 + 1)$. The line passes through $(1, 3)$ if and only if $3 = a + b$. Hence $b = 3 - a$, so $2(a^2 + 1) = b^2 = 9 - 6a + a^2$ and $a^2 + 6a - 7 = 0$. Thus $a = 1$ or $a = -7$. The two tangent lines are $y = x + 2$ and $y = -7x + 10$.

16. The distances from (x, y) to $(1, 1)$ and to the line $x + y = 0$ are, respectively,

$$\sqrt{(x-1)^2 +(y-1)^2}$$

and

$$\frac{|x+y|}{\sqrt{2}}$$

(the latter by Problem 14 in Chapter 2). Hence the parabola's equation sets these two distances equal. Squaring and multiplying gives $x^2 + y^2 - 2xy - 4x - 4y + 4 = 0$.

17. Let the chord be $y = mx + a$. At the points of intersection, $x^2 = 4py = 4pmx + 4pa$ so $x^2 - 4pmx - 4pa = 0$ and

$$x = 2pm \pm 2\sqrt{p^2 m^2 + pa}.$$

The x-coordinate of the midpoint is thus $2pm$. Because this doesn't depend on a, all such midpoints lie on the vertical line $x = 2pm$. Hence the locus is the half of this line that's "inside" the parabola; that is, above it if $p > 0$ and below it if $p < 0$.

Chapter 4

1. (a) $f(-3) = 42$.
    (b) $f(2) = 17$.
    (c) $f(0) = -3$.
    (d) $f(-\sqrt{7}) = 32$.
    (e) $f(a + 3) = 5a^2 + 30a + 42$.
    (f) $f(5t) = 125t^2 - 3$.

2. (a) $g(3) = 1/2$.
    (b) $g(-3) = 2$.
    (c) $g(1/3) = -1/2$.
    (d) $g(1/a) = (1 - a)/(a + 1)$.
    (e) $g(a + 1) = a/(a + 2)$.
    (f) $g(t - 1) = (t - 2)/t$.

3. (a) $\dfrac{[5(x+h)-3]-[5x-3]}{h} = \dfrac{5h}{h} = 5.$

(b) $\dfrac{(x+h)^2 - x^2}{h} = \dfrac{2xh+h^2}{h} = 2x+h.$

(c) $\dfrac{\dfrac{1}{x+h} - \dfrac{1}{x}}{h} = \dfrac{x-(x+h)}{hx(x+h)} = -\dfrac{1}{x(x+h)}.$

4. $f(x) = x^3 - 3x^2 + 4x - 2$, so $f(1) = 0, f(2) = 2, f(3) = 10, f(0) = -2, f(-1) = -10$, and $f(-2) = -30.$

5. $f(x) = 2^x$, so $f(1) = 2, f(3) = 8, f(5) = 32, f(0) = 1$, and $f(-2) = 1/4.$

6. $f(2x) = 4(2x) - 3 = 8x - 3 = 8x - 6 + 3 = 2(4x - 3) + 3 = 2f(x) + 3.$

7. The domain of $f(x) = 1/(x-8)$ consists of all x except 8. The domain of $g(x) = x^3$ is $(-\infty, \infty)$. $h(x) = f(g(x)) = f(x^3) =$

$$\frac{1}{(x-2)(x^2+2x+4)}.$$

The domain of $h(x)$ consists of all x except 2.

8. (a) $\sqrt{x}$ if and only if $x \geq 0$; the domain is therefore $[0, \infty)$.

(b) For $-\sqrt{x}$ the domain is $(-\infty, 0]$.

(c) $x^2 \geq 0$ for all x, so the domain is $(-\infty, \infty)$.

(d) $x - 4 \geq 0$ if and only if $x \leq -2$ or $x \geq 2$; the domain therefore consists of $(-\infty, -2]$ and $[2, \infty)$.

(e) $x^2 - 4 \neq 0$ if and only if $x \neq -2$ and $x \neq 2$; the domain therefore consists of all x except -2 and 2.

(f) $x^2 + 4 \neq 0$ for all x; the domain is thus $(-\infty, \infty)$.

(g) $(x-1)(x+2) \geq 0$ if and only if $x \leq -2$ or $x \geq 1$; the domain therefore consists of $(-\infty, -2]$ and $[1, \infty)$.

(h) $(x-1)(x+2) > 0$ if and only if $x < -2$ or $x > 1$; the domain therefore consists of $(-\infty, -2)$ and $(1, \infty)$.

(i) $3 - 2x - x^2 = (3 + x)(1 - x) \geq 0$ if and only if $-3 \leq x \leq 1$; the domain is $[-3, 1]$.

(j) For $x = 2$, $x/(x - 2) = 2/0$ is undefined. For $x \neq 2$, $x/(x - 2) = [x(x - 2)]/(x - 2)^2 \geq 0$ if and only if $x(x - 2) \geq 0$; that is, $x \leq 0$ or $x \geq 2$. The domain therefore consists of $(-\infty, 0]$ and $(2, \infty)$.

9. $f(x) = x/(x - 1)$, so $f(0) = 0/{-1} = 0$; $f(1) = 1/0$ is undefined; $f(2) = 2/1 = 2$; $f(3) = 3/2$; $f(f(3)) = f(3/2) = (3/2)/(3/2 - 1) = (3/2)/(1/2) = 3$. For $x \neq 1$,

$$f(f(x)) = f\left(\frac{x}{x-1}\right) = \frac{\dfrac{x}{x-1}}{\dfrac{x}{x-1} - 1} = \frac{x}{x - (x-1)} = x.$$

10. $f(x) = 1/(1 - x)$, so $f(0) = 1/1 = 1$; $f(1) = 1/0$ is undefined; $f(2) = 1/{-1} = -1$;

$$f(f(2)) = f(-1) = \frac{1}{1 - (-1)} = \frac{1}{2};$$

$$f(f(f(2))) = f(f(-1)) = f\left(\frac{1}{2}\right) = \frac{1}{1 - \dfrac{1}{2}} = 2.$$

For $x \neq 0$ and $x \neq 1$,

$$f(f(x)) = f\left(\frac{1}{1-x}\right) = \frac{1}{1 - \dfrac{1}{1-x}} = \frac{1-x}{(1-x) - 1} = \frac{x-1}{x}$$

and

$$f(f(f(x))) = f\left(\frac{x-1}{x}\right) = \frac{1}{1 - \dfrac{x-1}{x}} = \frac{x}{x - (x-1)} = x.$$

11. $f(x) = 2^x$; $f(a)f(b) = f(a + b)$.

12. $f(g(x)) = f(cx + d) = a(cx + d) + b = acx + (ab + b)$;

$g(f(x)) = g(ax + b) = c(ax + b) + d = acx + (bc + d)$.

Hence $f(g(x))$ need not equal $g(f(x))$. For example, we can set $a = b = c = x = 0$ and $d = 1$.

13. (a) The three equations obtained from evaluating $f(x) = ax^2 + bx + c$, given that $f(0) = 3, f(1) = 2$, and $f(2) = 9$, are $c = 3$, $a + b + c = 2$, and $4a + 2b + c = 9$. Hence $a + b = 2 - c = -1$ and $2a + b = (9 - c)/2 = 3$, so $a = (2a + b) - (a + b) = 3 - (-1) = 4$ and $b = (a + b) - a = -1 - 4 = -5$. The coefficients are $a = 4$, $b = -5$, and $c = 3$.

(b) By completing the square, we have

$$ax^2 + bx + c = a\left(x + \frac{b}{2a}\right)^2 + \frac{4ac - b^2}{4a}.$$

If $a > 0$, then $ax^2 + bx + c \geq \dfrac{4ac - b^2}{4a}$ for all x.

If $a < 0$, then $ax^2 + bx + c \leq \dfrac{4ac - b^2}{4a}$ for all x.

14. If $2x^2 + 2xy + y^2 = 3$ then $y^2 + 2xy + (2x^2 - 3) = 0$, so

$$y = \frac{-2x \pm \sqrt{(2x)^2 - 4(2x^2 - 3)}}{2} = -x \pm \sqrt{3 - x^2}.$$

The desired equations are

$$y = -x + \sqrt{3 - x^2} \quad \text{and} \quad y = -x - \sqrt{3 - x^2}.$$

15.

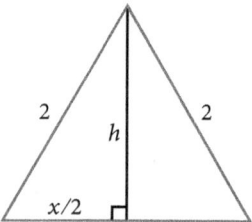

As shown in the figure, the height h satisfies

$$2^2 = h^2 + \left(\frac{x}{2}\right)^2,$$

so

$$h = \sqrt{4 - \frac{x^2}{4}} = \frac{1}{2}\sqrt{16 - x^2}.$$

Hence the area is

$$A = \frac{1}{2}xh = \frac{1}{2}x \cdot \frac{1}{2}\sqrt{16-x^2} = \frac{1}{4}x\sqrt{16-x^2}.$$

16.

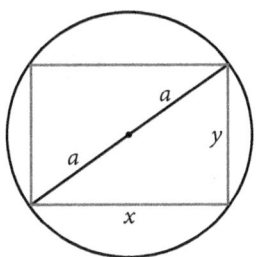

Let y be the length of the other side, as shown in the figure. Therefore

$$x^2 + y^2 = (2a)^2,$$

so

$$y = \sqrt{4a^2 - x^2}.$$

Hence the area is

$$A = xy = x\sqrt{4a^2 - x^2}.$$

17. $p = 4s$, so $s = p/4$ and $A = s^2 = p^2/16$.

18. $c = 2\pi r$, so $r = c/2\pi$ and $A = \pi r^2 = c^2/4\pi$.

19. $h = \frac{1}{2}\sqrt{3}b$.

20. Let the height be h. Then the volume is $V = \pi r^2 h$, so $h = V/\pi r^2$ and the area is $A = 2\pi r^2 + 2\pi rh = 2\pi r^2 + 2V/r$.

21. If the length of the other side is y feet, then the perimeter is $100 = 2x + 2y$, so $y = 50 - x$ and the area is $A = xy = 50x - x^2 = 625 - (x^2 - 50x + 625) = 625 - (x - 25)^2$. This area will be maximal when $(x - 25)^2$ is minimal; that is, for $x = 25$. The maximal area is 625 feet^2 and the pen is 25 feet by 25 feet.

22. (a) $(-\infty, \infty)$.
    (b) All $x \neq 3/2$.
    (c) $[2/3, \infty)$.
    (d) $(-\infty, 5/3]$.

(e) All $x \neq 3, -3$.

(f) $(-\infty, \infty)$.

(g) $[-3/2, 3/2]$.

(h) $(-3, \infty)$.

(i) $(-\infty, \infty)$.

23.
$$f\left(\frac{x_1 + x_2}{2}\right) = a\frac{x_1 + x_2}{2} + b$$
$$= \frac{ax_1 + ax_2 + 2b}{2}$$
$$= \frac{(ax_1 + b) + (ax_2 + b)}{2}$$
$$= \frac{f(x_1) + f(x_2)}{2}.$$

24. $g(x) = x^3$, so $g(f(x)) = g(\sqrt[3]{x}) = (\sqrt[3]{x})^3 = x$.

25. If the height is h, then $A = 2\pi r^2 + 2\pi rh = 2\pi r(r + h)$, so $h = (A/2\pi r) - r$. Therefore, the volume is

$$V = \pi r^2 h = \pi r^2\left(\frac{A}{2\pi r} - r\right) = \frac{A}{2}r - \pi r^3.$$

Chapter 5

1. (a) $y = x^2 + x - 2 = (x - 1)(x + 2)$. y is positive when $x < -2$ or $x > 1$, negative when $-2 < x < 1$, and zero when $x = -2$ or $x = 1$. y is large when x is large.

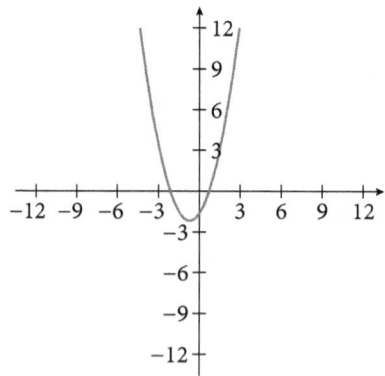

(b) $y = x^3 - 3x^2 + 2x = x(x-1)(x-2)$. y is positive when $0 < x < 1$ or $x > 2$, negative when $x < 0$ or $1 < x < 2$, and zero when $x = 0, 1,$ or 2. y is large when x is large.

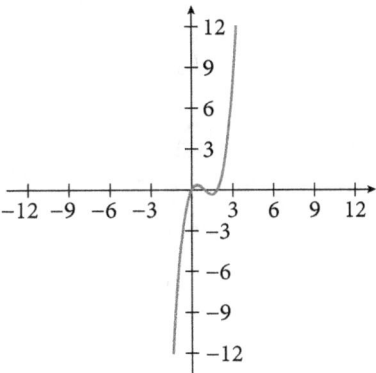

(c) $y = (1-x)(2-x)(3-x)$. y is positive when $x < 1$ or $2 < x < 3$, negative when $1 < x < 2$ or $x > 3$, and zero when $x = 1, 2,$ or 3. y is large when x is large.

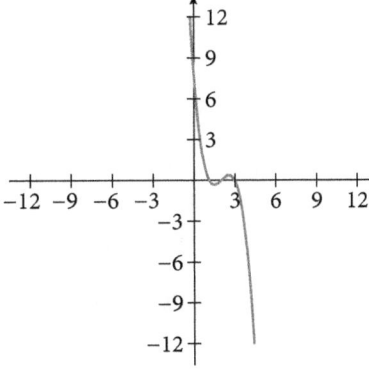

(d) $y = x^4 - x^2 = x^2(x - 1)(x + 1)$. y is positive when $x < -1$ or $x > 1$, negative when $-1 < x < 0$ or $0 < x < 1$, and zero when $x = -1, 0,$ or 1. y is large when x is large.

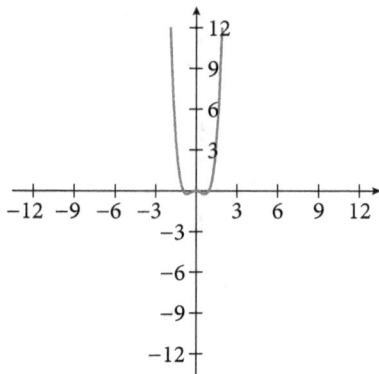

(e) $y = x^4 - 5x^2 + 4 = (x^2 - 1)(x^2 - 4) = (x + 1)(x - 1)(x + 2)(x - 2)$. y is positive when $x < -2, -1 < x < 1,$ or $x > 2$, negative when $-2 < x < -1$ or $1 < x < 2$, and zero when $x = -2, -1, 1,$ or 2. y is large when x is large.

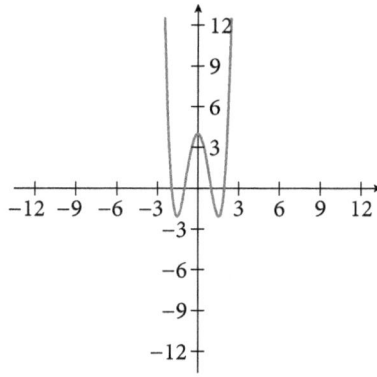

2.

(a) $y = \dfrac{1}{x^2}$

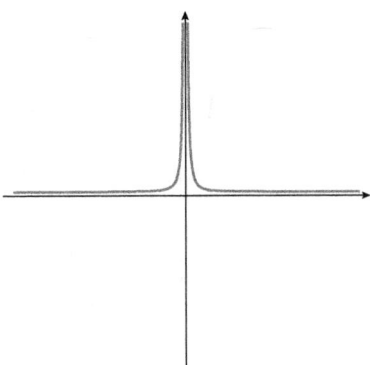

(b) $y = \dfrac{1}{x^3}$

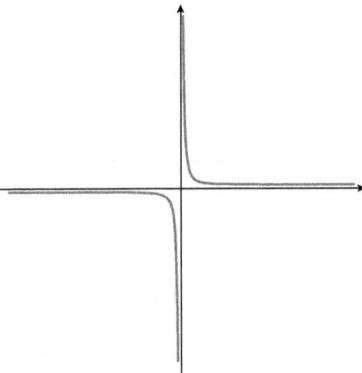

(c) $y = x^2 + \dfrac{1}{x}$

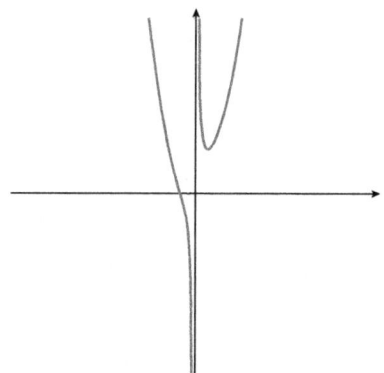

(d) $y = x^2 + \dfrac{1}{x^2}$

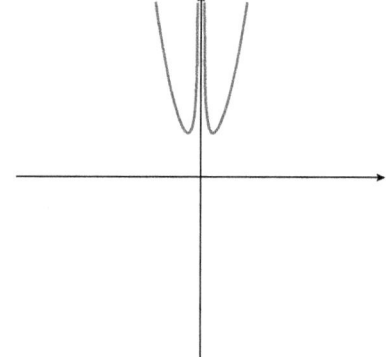

(e) $y = \dfrac{1}{x^2 + 1}$

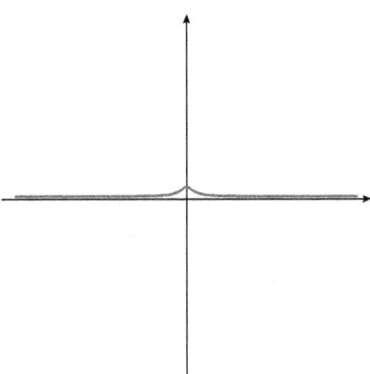

(f) $y = \dfrac{x^2}{x^2 + 1}$

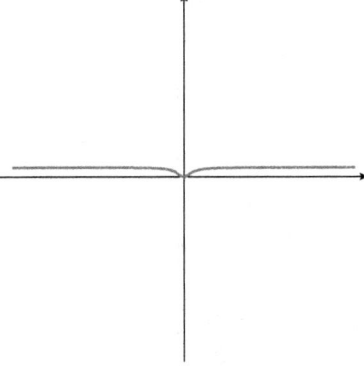

(g) $y = \dfrac{1}{x^2 - 1}$

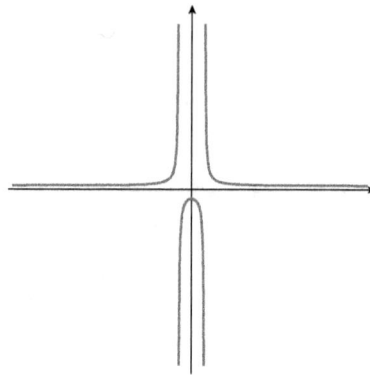

(h) $y = \dfrac{x}{x^2 - 1}$

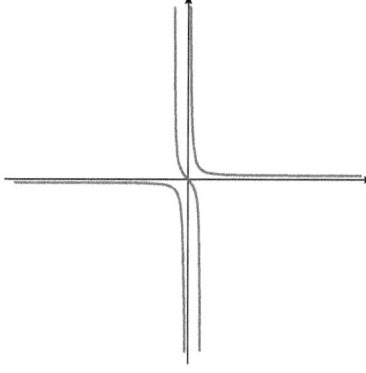

(i) $y = \dfrac{x^2}{x^2 - 1}$

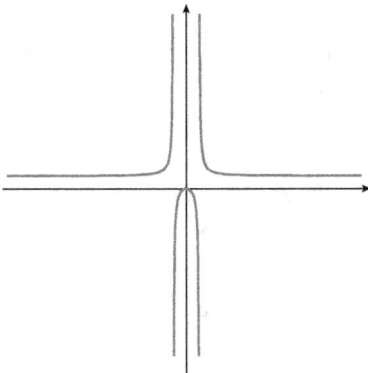

3. (a) $y = \sqrt{(x-1)(3-x)}$

$\qquad = \sqrt{-x^2 + 4x - 3}$

$\qquad = \sqrt{1 - (x^2 - 4x + 4)}$

$\qquad = \sqrt{1 - (x-2)^2}$,

so $(x - 2)^2 + y^2 = 1$. The graph is the upper half of the unit circle centered at (2, 0).

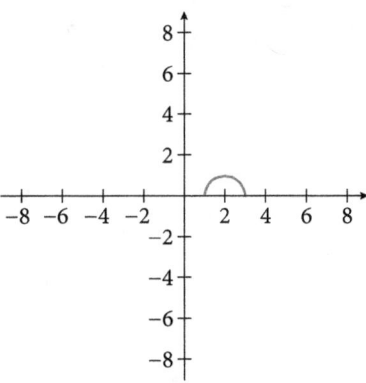

(b) $y = \dfrac{1}{\sqrt{(x-1)(3-x)}}$. y is defined when $(x - 1)(3 - x) > 0$; that is,

for $1 < x < 3$. It's always positive and is large for x near 1 or 3.

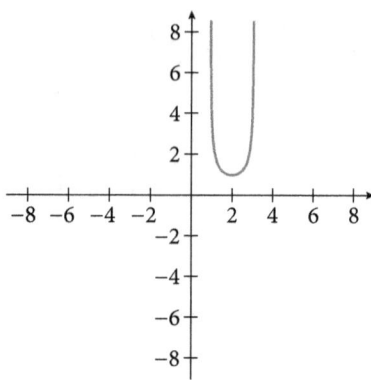

(c) $y = \dfrac{1}{\sqrt{x-1}}$. y is defined for $x > 1$. It's large for x near (but greater

than) 1 and is small when x is large.

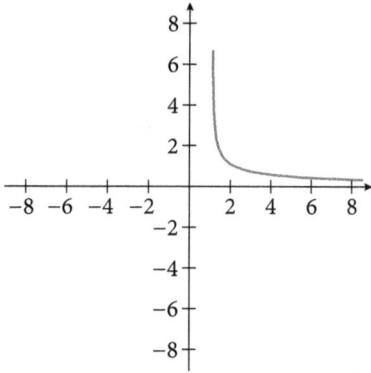

(d) $y = \sqrt{\dfrac{x}{3-x}}$. y is defined when $x/(3 - x) \geq 0$.

But $\dfrac{x}{3-x} = \dfrac{x(3-x)}{(3-x)^2}$. The denominator is positive for $x \neq 3$, so $x/(3 -$

$x) \geq 0$ if and only if $x \neq 3$ and $x(3 - x) \geq 0$; that is, for $0 \leq x < 3$. y
is zero for $x = 0$ and positive for $0 < x < 3$. y is large for x near (but
less than) 3.

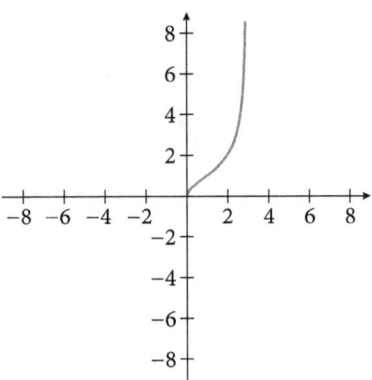

(e) $y = \sqrt{\dfrac{4-x}{x-2}} = \sqrt{\dfrac{(4-x)(x-2)}{(x-2)^2}}$. y is defined when $x \neq 2$ and $(4 - x)$

$(x - 2) \geq 0$; that is, for $2 < x \leq 4$. y is zero for $x = 4$ and positive for
$2 < x \leq 4$. y is large for x near (but greater than) 2.

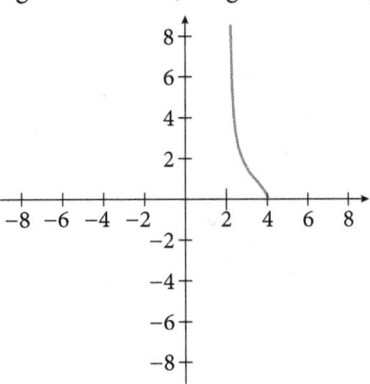

(f) $y = \sqrt{\dfrac{x-4}{x-2}} = \sqrt{\dfrac{(x-2)(x-4)}{(x-2)^2}}$. y is defined when $x \neq 2$ and $(x-2)$

$(x-4) \geq 0$; that is, for $x < 2$ and for $x \geq 4$. For large x, $(x-4)/(x-2)$ is near 1 so y is near 1; the line $y = 1$ is an asymptote. y is large for x near (but less than) 2; that is, the line $x = 2$ is an asymptote. y is zero when $x = 4$ and positive otherwise.

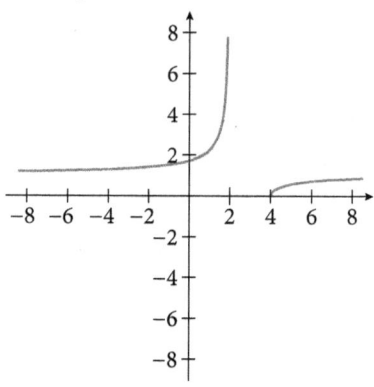

4. (a) $y = \dfrac{|x|}{x} = \begin{cases} 1, & \text{if } x > 0 \\ -1, & \text{if } x < 0 \end{cases}$ and is undefined for $x = 0$.

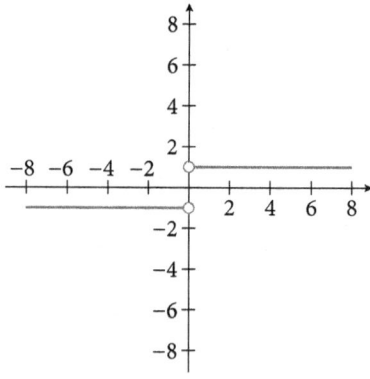

(b) $y = |2x + 3| = \begin{cases} 2x + 3, & \text{if } x \geq -3/2 \\ -(2x + 3), & \text{if } x < -3/2 \end{cases}$

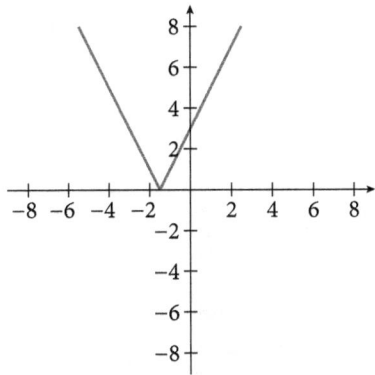

(c) $y = x + |x| = \begin{cases} 2x, & \text{if } x \geq 0 \\ 0, & \text{if } x < 0 \end{cases}$

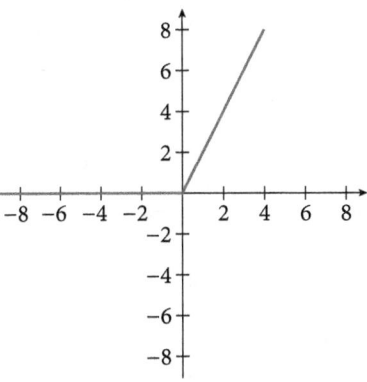

(d) $y = 2x + |x| = \begin{cases} 3x, & \text{if } x \geq 0 \\ x, & \text{if } x < 0 \end{cases}$

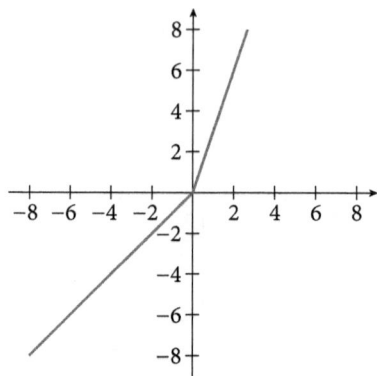

(e) $y = x - |x| = \begin{cases} 0, & \text{if } x \geq 0 \\ 2x, & \text{if } x < 0 \end{cases}$

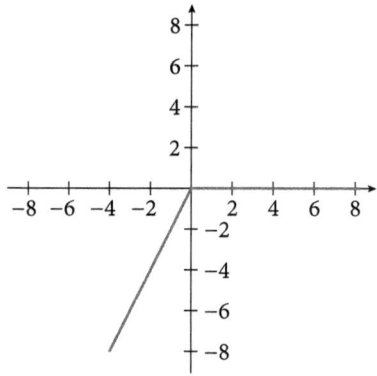

(f) $y = 1 + x - |x| = \begin{cases} 1, & \text{if } x \geq 0 \\ 1 + 2x, & \text{if } x < 0 \end{cases}$

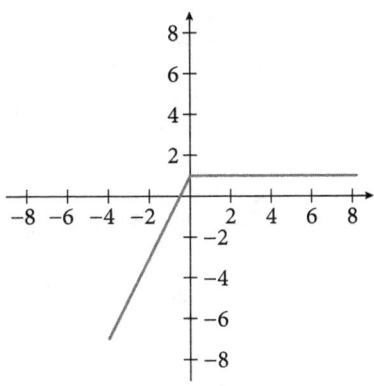

(g) $y = |x^2 - 1| = \begin{cases} x^2 - 1, & \text{if } x \leq -1 \\ 1 - x^2, & \text{if } -1 \leq x \leq 1 \\ x^2 - 1, & \text{if } x \geq 1 \end{cases}$

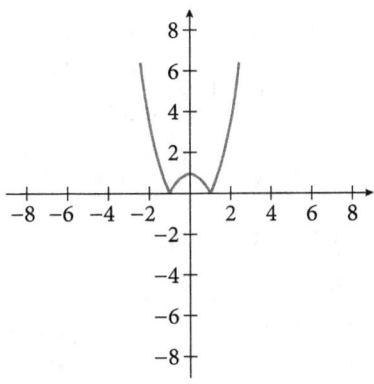

5. (a) No; $f(x) = x/x$ and $g(x) = 1$ aren't equal because x/x isn't defined for $x = 0$.

    (b) Yes; $f(x) = x^2 - 1$ and $g(x) = (x + 1)(x - 1)$ are equal.

    (c) No; $f(x) = x$ and $g(x) = \sqrt{x^2} = |x|$ don't agree for $x < 0$.

    (d) No; $f(x) = x$ and $g(x) = (\sqrt{x})^2$ don't agree for $x < 0$, where $g(x)$ isn't defined.

6. Let $f(x) = (x - 1)(x - 2) \cdots (x - n)$. Then f is a polynomial of degree n with n zeros, namely, 1, 2,..., n. For n even, let $g(x) = x^n + 1$. For all x, $g(x) = (x^{n/2})^2 + 1 > 0$, so g has no zeros. For n odd, let $g(x) = x^n$. Then $g(x) = 0$ if and only if $x = 0$; thus, g has only one zero.

7. (a) Odd: $f(-x) = (-x)^3 = -x^3 = -f(x)$.

    (b) Even: $f(-x) = (-x)((-x)^3 + (-x)) = -x(-x^3 - x) = x(x^3 + x) = f(x)$.

    (c) Even: $f(-x) = |-x| = |x| = f(x)$.

    (d) Odd: $f(-x) = -x + (1/-x) = -x - (1/x) = -(x + 1/x) = -f(x)$.

    (e) Neither: $f(1) = 2$, whereas f(-1) = 0.

    (f) Odd: Because $x^2 + 1 \neq 0$, $f(x) = \dfrac{x(x^2 + 1)}{x^2 + 1} = x$; hence $f(-x) = -x = -f(x)$.

    (g) Neither: $f(1) = 2$, whereas $f(-1) = 0$.

    (h) Neither: $f(1) = 2$, whereas $f(-1) = 0$.

8. The graph of an even function is symmetric with respect to the y-axis, meaning that its graph remains unchanged after reflection about the y-axis. The graph of an odd function has rotational symmetry with respect to the origin, meaning that its graph remains unchanged after rotation of 180 degrees about the origin.

9. (a) The product of two even functions is even. If f and g are even, then $(fg)(-x) = f(-x)g(-x) = f(x)g(x) = (fg)(x)$.

(b) The product of two odd functions is even. If f and g are odd, then $(fg)(-x) = f(-x)g(-x) = (-f(x))(-g(x)) = f(x)g(x) = (fg)(x)$.

(c) The product of an even function and an odd function is odd. If f is odd and g is even, then $(fg)(-x) = f(-x)g(-x) = (-f(x))g(x) = -f(x) g(x) = -(fg)(x)$.

10. Consider the polynomial $f(x) = a(x - 2)(x - 3) + b(x - 1)(x - 3) + c(x - 1)(x - 2)$. We have $f(1) = 2a$, $f(2) = -b$, and $f(3) = 2c$. So, letting $a = \pi/2$, $b = -\sqrt{3}$, and $c = 550/2 = 275$, we obtain the desired polynomial $f(x) = \pi/2(x - 2)(x - 3) - \sqrt{3}(x - 1)(x - 3) + 275(x - 1)(x - 2)$.

11. (a) $y = [x]$

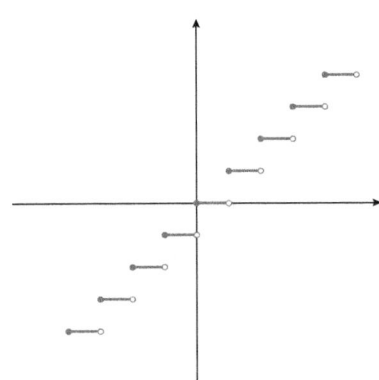

(b) $y = x - [x]$

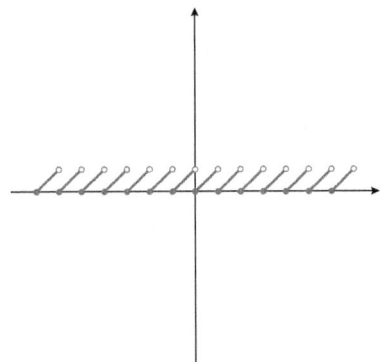

(c) $y = \sqrt{x - [x]}$

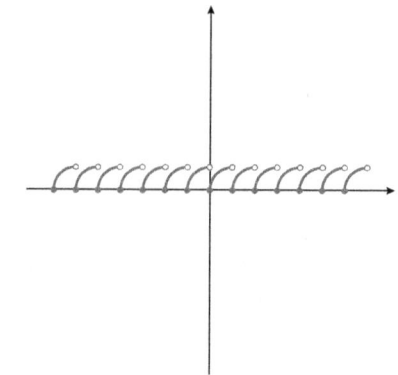

(d) $y = [x] + \sqrt{x - [x]}$

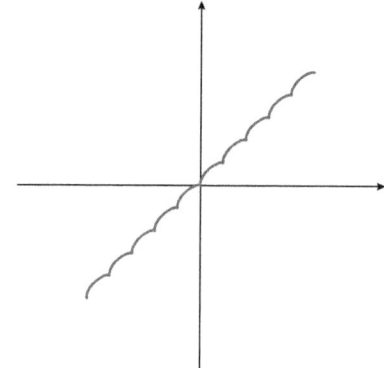

Chapter 6

1. (a) $15° = 15 \times \pi/180 = \pi/12$ radians.
   (b) $150° = 150 \times \pi/180 = 5\pi/6$ radians.
   (c) $1500° = 50\pi/6$ radians.
   (d) $-36° = -\pi/5$ radians.
   (e) $7° = 7\pi/180$ radians.
   (f) $-110° = -11\pi/18$ radians.

2. (a) $\pi/15 = \pi/15 \times 180/\pi = 12°$.
   (b) $\pi/45 = \pi/45 \times 180/\pi = 4°$.
   (c) $-\pi/36 = -5°$.
   (d) $-3 = -540°/\pi$.

(e) $\pi^2 = 180\pi°$.

(f) $30 = 5400°/\pi$.

3. (a) Note that $-120° = -2\pi/3$ is one-third of the way from $-\pi/2$ to $-\pi$, so the endpoint on the unit circle is in the third quadrant with coordinates $(-\frac{1}{2}, -\sqrt{3}/2)$, so $\cos -120° = -\frac{1}{2}$.

(b) Because $\sin\theta$ is periodic with period $2\pi = 360°$,

$$\sin 780° = \sin(2 \cdot 360° + 60°) = \sin 60° = \sin\frac{\pi}{3} = \frac{\sqrt{3}}{2}.$$

(c) Because $\sin\theta$ is periodic with period 2π,

$$\sin\frac{17\pi}{3} = \sin\left(2 \cdot 2\pi + \frac{5\pi}{3}\right) = \sin\frac{5\pi}{3} = -\frac{\sqrt{3}}{2}.$$

(d) Because $\cos\theta$ is periodic with period 2π,

$$\cos\frac{-15\pi}{4} = \cos\left(\frac{\pi}{4} - 2 \cdot 2\pi\right) = \cos\frac{\pi}{4} = \frac{\sqrt{2}}{2}.$$

(e) Because $\sin\theta$ is periodic with period 2π,

$$\sin\frac{19\pi}{6} = \sin\left(2\pi + \frac{7\pi}{6}\right) = \sin\frac{7\pi}{6}.$$

Note that $7\pi/6$ is one-third of the way from π to $3\pi/2$, so the endpoint on the unit circle is in the third quadrant with coordinates $(-\sqrt{3}/2, -\frac{1}{2})$, so $\sin 19\pi/6 = -\frac{1}{2}$.

(f) Because $\cos\theta$ is periodic with period 2π,

$$\cos\frac{99\pi}{4} = \cos\left(12 \cdot 2\pi + \frac{3\pi}{4}\right) = \cos\frac{3\pi}{4} = -\frac{\sqrt{2}}{2}.$$

4. (a) $\sin 500\pi$ is zero.

(b) $\cos 7$ is positive.

(c) $\sin 901°$ is negative.

(d) $\cos 2^4$ is positive.

5. $\sin 3\theta = \sin(2\theta + \theta)$

$$= \sin 2\theta \cos\theta + \cos 2\theta \sin\theta$$

$$= (2\sin\theta \cos\theta)\cos\theta + (\cos^2\theta - \sin^2\theta)\sin\theta$$

$$= 2\sin\theta \cos^2\theta + \sin\theta \cos^2\theta - \sin^3\theta$$

$$= 3\sin\theta \cos^2\theta - \sin^3\theta$$

$$= 3\sin\theta(1 - \sin^2\theta) - \sin^3\theta$$

$$= 3\sin\theta - 4\sin^3\theta.$$

6. $\cos 3\theta = \cos(2\theta + \theta)$

$$= \cos 2\theta \cos\theta - \sin 2\theta \sin\theta$$

$$= (\cos^2\theta - \sin^2\theta)\cos\theta - (2\sin\theta \cos\theta)\sin\theta$$

$$= \cos^3\theta - 3\sin^2\theta \cos\theta$$

$$= \cos^3\theta - 3(1 - \cos^2\theta)\cos\theta$$

$$= 4\cos^3\theta - 3\cos\theta.$$

7. The periodicity of sine and cosine lets us reduce the angle to one in the range $[-\pi, \pi]$. To reduce the range further, we'll use the following identities:

$$\sin(\theta + \pi) = -\sin\theta,$$

$$\cos(\theta + \pi) = -\cos\theta,$$

These identities follow by letting $\theta = \pi$ in (3) and (4). By using them we can reduce the angle to one in the range $[-\pi/2, \pi/2]$. Finally, equations (1) and (2) let us replace the angle by one in the first quadrant.

(a) $\sin\left(\dfrac{9\pi}{2}\right) = \sin\left(\dfrac{\pi}{2} + 2 \cdot 2\pi\right) = \sin\left(\dfrac{\pi}{2}\right).$

(b) $\sin 7\pi = \sin(\pi + 3 \cdot 2\pi) = \sin \pi = -\sin 0.$

(c) $\sin\left(-\dfrac{7\pi}{3}\right) = \sin\left(-\dfrac{\pi}{3} - 2\pi\right) = \sin\left(-\dfrac{\pi}{3}\right) = -\sin\dfrac{\pi}{3}.$

(d) $\sin\left(-\dfrac{8\pi}{3}\right) = \sin\left(-\dfrac{2\pi}{3} - 2\pi\right) = \sin\left(-\dfrac{2\pi}{3}\right)$

$$= \sin\left(\dfrac{\pi}{3} - \pi\right) = -\sin\dfrac{\pi}{3}.$$

(e) $\cos 10\pi = \cos(0 + 5 \cdot 2\pi) = \cos 0.$

(f) $\cos\left(\dfrac{9\pi}{4}\right) = \cos\left(\dfrac{\pi}{4} + 2\pi\right) = \cos\dfrac{\pi}{4}.$

(g) $\cos\left(-\dfrac{6\pi}{5}\right) = \cos\left(-\dfrac{\pi}{5} - \pi\right) = -\cos\left(-\dfrac{\pi}{5}\right) = -\cos\dfrac{\pi}{5}.$

(h) $\sin\left(-\dfrac{11\pi}{2}\right) = \sin\left(\dfrac{\pi}{2} - 3 \cdot 2\pi\right) = \sin\dfrac{\pi}{2}.$

(i) $\cos\left(\dfrac{11\pi}{3}\right) = \cos\left(-\dfrac{\pi}{3} + 2 \cdot 2\pi\right) = \cos\left(-\dfrac{\pi}{3}\right) = \cos\dfrac{\pi}{3}.$

8. $\sin(\pi - \theta) = y = \sin\theta$ and $\cos(\pi - \theta) = -x = -\cos\theta.$

9. $\sin(\pi/2 - \theta) = x = \cos\theta$ and $\cos(\pi/2 - \theta) = y = \sin\theta.$

10. Because 15° is in the first quadrant, both sin 15° and cos 15° are positive. So

$$\sin 15° = \sqrt{\dfrac{1}{2}(1 - \cos 30°)} = \sqrt{\dfrac{1}{2}\left(1 - \dfrac{1}{2}\sqrt{3}\right)} = \dfrac{1}{2}\sqrt{2 - \sqrt{3}}$$

and

$$\cos 15° = \sqrt{\dfrac{1}{2}(1 + \cos 30°)} = \sqrt{\dfrac{1}{2}\left(1 + \dfrac{1}{2}\sqrt{3}\right)} = \dfrac{1}{2}\sqrt{2 + \sqrt{3}}.$$

11. (a) $\cos\dfrac{\pi}{4} = \sqrt{\dfrac{1}{2}(1+\cos\dfrac{\pi}{2})} = \sqrt{\dfrac{1}{2}(1+0)} = \dfrac{1}{2}\sqrt{2}.$

(b) Because $3\pi/4$ is in the second quadrant, $\cos 3\pi/4$ is negative. Therefore

$$\cos\dfrac{3\pi}{4} = -\sqrt{\dfrac{1}{2}(1+\cos\dfrac{3\pi}{2})} = -\sqrt{\dfrac{1}{2}(1+0)} = -\dfrac{1}{2}\sqrt{2}.$$

12. $\cos\dfrac{5\pi}{4} = \cos\left(\pi + \dfrac{\pi}{4}\right)$

$$= \cos\pi \cos\dfrac{\pi}{4} - \sin\pi \sin\dfrac{\pi}{4}$$

$$= (-1)\left(\dfrac{1}{2}\sqrt{2}\right) - 0\left(\dfrac{1}{2}\sqrt{2}\right)$$

$$= -\dfrac{1}{2}\sqrt{2}.$$

13. $\sin(\theta + \varphi) = \sin\left(\dfrac{\pi}{6} + \dfrac{\pi}{3}\right) = \sin\dfrac{\pi}{2} = 1.$

$\sin(\theta + \varphi) = \sin\theta\cos\varphi + \cos\theta\sin\varphi$

$$= \sin\dfrac{\pi}{6}\cos\dfrac{\pi}{3} + \cos\dfrac{\pi}{6}\sin\dfrac{\pi}{3}$$

$$= \dfrac{1}{2}\cdot\dfrac{1}{2} + \dfrac{1}{2}\sqrt{3}\cdot\dfrac{1}{2}\sqrt{3}$$

$$= \dfrac{1}{4} + \dfrac{3}{4} = 1.$$

The results are equal.

14. $\sin\dfrac{5\pi}{12} = \sin(\dfrac{\pi}{4} + \dfrac{\pi}{6})$

$$= \sin\dfrac{\pi}{4}\cos\dfrac{\pi}{6} + \cos\dfrac{\pi}{4}\sin\dfrac{\pi}{6}$$

$$= \dfrac{1}{2}\sqrt{2}\cdot\dfrac{1}{2}\sqrt{3} + \dfrac{1}{2}\sqrt{2}\cdot\dfrac{1}{2}$$

$$= \dfrac{1}{4}(\sqrt{6} + \sqrt{2}).$$

15.

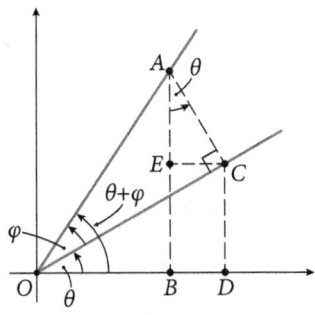

$$\cos(\theta+\varphi) = \frac{OB}{OA} = \frac{OD - BD}{OA} = \frac{OD - EC}{OA}$$

$$= \frac{OD}{OA} - \frac{EC}{OA}$$

$$= \frac{OD}{OC} \cdot \frac{OC}{OA} - \frac{EC}{AC} \cdot \frac{AC}{OA}$$

$$= \cos\theta \cos\varphi - \sin\theta \sin\varphi.$$

16. To prove the law of sines, add the triangle's height h:

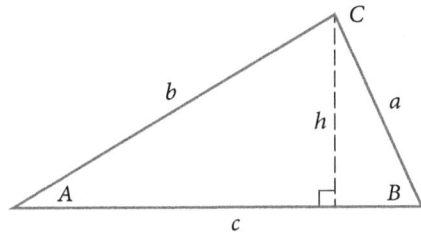

For the two resulting right triangles, we have

$$\sin A = h/b \qquad \text{and} \qquad \sin B = h/a.$$

Rearranging terms yields

$$h = b \sin A \qquad \text{and} \qquad h = a \sin B.$$

Equate these two expressions to h to get $b \sin A = a \sin B$ or

$$\frac{\sin A}{a} = \frac{\sin B}{b}.$$

The remaining part of the law of sines is proved in the same way.

17.

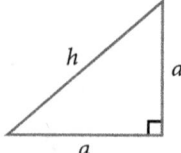

If the legs have length a, as shown in the figure, then
$$h = \sqrt{a^2 + a^2} = \sqrt{2}a,$$

so

$$a = \frac{h}{\sqrt{2}}.$$

Hence the area is

$$A = \frac{1}{2}a^2 = \frac{1}{4}h^2.$$

Index

right triangle 69
sec (secant) function 82
sin (sine) function 69, 73
subtended angles 71
subtraction formulas 76
tan (tangent) function 82
unit circle 71, 73

U

unit circle 71, 73
unit distance 2

V

vertex (parabola) 33

X

x-axis 7
x-coordinate 7
x-intercept 25
xy-plane. *See* coordinate plane

Y

y-axis 7
y-coordinate 7
y-intercept 21

Z

zero point 2
zeros of a function 57

www.ingramcontent.com/pod-product-compliance
Lightning Source LLC
Chambersburg PA
CBHW070342220526
45467CB00001B/228